Wastewater Engineering and Treatment 4

Wastewater Engineering and Treatment 4

Micropollutants, Lagoons, Thickening, Mud and Odor Treatment

Kader Gaid

iSTE

WILEY

First published 2026 in Great Britain and the United States by ISTE Ltd and John Wiley & Sons, Inc.

ISTE Ltd
27-37 St George's Road
London SW19 4EU
UK

www.iste.co.uk

John Wiley & Sons, Inc.
111 River Street
Hoboken, NJ 07030
USA

www.wiley.com

Library of Congress Control Number: 2026938343

British Library Cataloguing-in-Publication Data
A CIP record for this book is available from the British Library
ISBN 978-1-83669-058-0

The manufacturer's authorized representative according to the EU General Product Safety Regulation is Wiley-VCH GmbH, Boschstr. 12, 69469 Weinheim, Germany, e-mail: Product_Safety@wiley.com.

Contents

1

Biological Removal of Micropollutants

1.1. Introduction

Wastewater treatment plants treat wastewater from a wide variety of sources including domestic, industrial, commercial or healthcare activities. The effluents contain numerous chemical substances, some of which are classified as "micropollutants" because they are present at very low concentrations (nanograms to micrograms per liter). Micropollutants are mineral or organic molecules originating from the chemical, pharmaceutical or agricultural industries, which, beyond a certain threshold, can have negative impacts on ecosystems and the environment. However, due to these human activities, micropollutants end up dispersed in aquatic environments as well as in the various environmental compartments (Kümmerer 2001).

Even at very low concentrations, some substances can significantly disrupt ecosystems, for example, by causing endocrine-disrupting effects (hormonal system dysfunction), or, in the case of antibiotics, by playing a major role in the spread of antibiotic resistance (Stalder 2012). Controlling the release of micropollutants into the environment has become a major issue for resource conservation and public health.

European regulations, particularly through directives 2000/60/EC and 2013/39/EU, define a list of 45 priority substances for which environmental quality standards must be met. These substances include trace metal elements (TMEs) and organic trace compounds (OTCs). OTCs are organic compounds present in very small quantities, but which pose a significant ecotoxicological risk. They are also called xenobiotics, organic micropollutants or, for some, persistent organic pollutants (POPs). OTCs include various families of compounds, such as polycyclic aromatic hydrocarbons (PAHs), polychlorinated biphenyls (PCBs), pesticides, dioxins, furans, detergent derivatives, plasticizers and natural and synthetic hormones. Pharmaceutical residues are also classified as organic micropollutants, but they are not currently regulated, despite their proven presence and impact on living organisms. Thus, a distinction is made between regulated and emerging pollutants.

1.2. Micropollutant origin

1.2.1. *Trace metal elements (TMEs)*

Trace elements are defined as elements present in the Earth's crust at less than one part per thousand. This concept should be distinguished from that of a microelement, which characterizes an element whose average dry matter concentration in living organisms is less than 100 $mg.kg^{-1}$. Certain major elements of the Earth's crust act like trace elements in their interactions with living organisms, such as iron and manganese (ADEME 1995). The metal functional group characterizes mineral elements that, in ionic form, exhibit a high affinity for the hydroxyl groups of water, with which they bind. The term "heavy metals" is frequently used to refer to mineral micropollutants. However, it encompasses both metals and metalloids, as well as elements with a density greater than 6 $g.mL^{-1}$ (ADEME 1995).

TMEs therefore include:

– trace elements (metals or non-metals) essential in small doses for living organisms: copper (Cu), zinc (Zn), boron (B), manganese (Mn), iron (Fe), nickel (Ni), vanadium (V), molybdenum (Mo), chromium (Cr), tin (Sn), arsenic (As), etc.;

– heavy metals for which no positive role in biological activity has been identified: mercury (Hg), lead (Pb) and cadmium (Cd).

Many metals, such as cadmium, chromium, copper, iron, lead, manganese, mercury, nickel and zinc, are naturally present in trace amounts in water. Most of these elements (cadmium, lead, mercury, nickel) are classified as priority pollutants in the Water Framework Directive 2000/60/EC from October 23, 2000. However, most of them are also essential elements for biological growth, and insufficient quantities can limit the growth of certain organisms, such as algae. Conversely, in excessive quantities and depending on their chemical forms, their toxicity can pose health problems and limit the use of water resources. It is therefore necessary to measure and control the concentrations of these substances.

Urban wastewater contains metal micropollutants in concentrations far exceeding those typically found in clean water (drinking water, groundwater, surface water). These pollutants originate from both domestic use and runoff from impermeable surfaces exposed to atmospheric pollution (rainwater, wash water) (ADEME 1995; Karvelas et al. 2003). The most common metallic micropollutants found in wastewater are iron, zinc, manganese, copper and nickel. These are followed by chromium, lead and cadmium (Lassachagne 2014).

1.2.2. *Organic trace compounds (OTCs)*

Organic trace compounds are organic compounds present in very small quantities, but which pose a significant ecotoxicological risk (Martín 2012a). They are also called xenobiotics, organic micropollutants or persistent organic pollutants (POPs). OTCs include different families of compounds such as PAHs (polycyclic aromatic hydrocarbons), PCBs (polychlorinated biphenyls), pesticides, dioxins, furans, detergent derivatives, plasticizers, natural and synthetic hormones, etc., as well as drug residues, otherwise known as pharmaceutical compounds.

For over 50 years, the production of organic chemical compounds for industrial and domestic uses has been steadily increasing (Rogers 1996). In addition to pesticides, residues of medicines and pharmaceutical substances are being studied with increasing frequency (Hirsch et al. 1998). They are considered emerging pollutants, residues from the use of chemicals present in the environment, not yet regulated but a topic with much scientific and public attention (Petrovic et al. 2003). This new concern is linked to two major phenomena: firstly, the emergence of modern analytical tools capable of measuring drug molecule concentrations at increasingly lower levels and, secondly, the greater scientific evidence of biological effects even at very low doses (Halling-Sørensen et al. 1998; Hirsch et al. 1998; Beausse 2004). Traces of these residues can subsequently be detected and quantified in different aquatic water cycle compartments: urban wastewater (Lindberg et al. 2005), the inlet and outlet effluents of wastewater treatment plants (Miao et al. 2002), etc.

Today, more than 100,000 chemical substances are listed in Europe, of which more than 30,000 are produced in quantities exceeding one ton. Among these, pharmaceutical compounds have been a major concern for several years and have also been identified as emerging environmental contaminants. Medicine is making significant progress thanks to medications with constantly evolving active substances that allow for the treatment of an ever-increasing number of human and animal diseases. Pharmaceutical compounds from human and animal medicines represent more than 4,000 molecules. After being consumed, the active compounds are often only partially metabolized and are then excreted via urine or feces. Thus, hospital, domestic or industrial wastewater sent to treatment plants is one of the largest sources (Hirsch et al. 1998; Beausse 2004; Mullot 2009; Deblonde et al. 2011; Verlicchi et al. 2012a).

1.2.2.1. *Pesticides*

Agriculture is by far the largest consumer of pesticides; the majority of active substances released into the environment originate from agriculture to combat pests and weeds. There are three main categories of biocides: insecticides, fungicides and herbicides.

Insecticides can affect the nervous system (neurotransmission), respiration or molting. Fungicides act selectively on respiration, biosynthesis (sterols, RNA, melanin), growth or cell permeability. Finally, herbicides can have a detrimental impact on photosynthesis, lipid and amino acid synthesis, cell division or growth. The most common pesticide families are listed as follows:

– Organophosphates (chlorpyrifos, malathion) have an insecticidal effect by blocking acetylcholinesterase (disrupting the transmission of nerve impulses).

– Organochlorines (DDT, chlordecone, lindane) are insecticides that act on the nervous system by disrupting nerve impulses at the synapses. These compounds are now all banned in France but are very persistent in the environment (the chlordecone problem in the French West Indies, for example).

– Carbamates (aldicarb, carbofuran) act on the nervous system in the same way as organophosphates.

– Triazines (atrazine, simazine) have a herbicidal effect by blocking electron transport and light energy.

– Phenylureas (diuron, isoproturon) have a mode of action comparable to that of triazines.

– Chloroacetanilides (metolachlor, acetochlor) act on plants by inhibiting elongases, enzymes involved in lipid synthesis (long-chain fatty acids).

– Aryloxy acids (2,4-D, mecoprop, dichlorprop) disrupt the regulation of auxin IAA (indole-3-acetic acid), a phytohormone essential for plant growth.

– Sulfonylureas (chlorsulfuron, nicosulfuron) act on the synthesis of branched-chain amino acids (valine, leucine, isoleucine) via the inhibition of acetolactate synthetase.

Agricultural activity is not the only source of pesticide emissions. Secondary sources include the maintenance of public spaces, golf courses, use by private individuals, annual mosquito control in wetlands, the veterinary sector, road maintenance, etc.

Main families	Compounds
Drug molecules	Anxiolytics, analgesics, anti-inflammatories, antibiotics, lipid-lowering drugs, antihypertensives, antiepileptics, anticancer drugs, iodinated contrast agents, etc.
Cosmetic and personal hygiene products	Musks (galaxolide, tonalide), triclosan, etc.
Endocrine disruptors	Hormones, alkylphenols, BPA, phthalates, flame retardants, etc.
Illegal drugs	Cocaine and metabolites, THC, amphetamines, etc.
Pesticides	Insecticides, herbicides
Priority substances (Annex X of the WFD) and hazardous substances (RSDE circular 050109)	PAHs, dichloromethane
Disinfection byproducts	Trihalomethanes, haloacetic acids

Table 1.1. *Main families and organic compounds*

1.2.2.2. *Pharmaceutical products and industrial waste*

France is one of the world's largest consumers of pharmaceuticals. More than 3,000 pharmaceutical molecules are commonly used in medicines, and 300 veterinary medicines are currently available on the French market. The biologically active substances contained in each pharmaceutical product have a wide diversity of chemical structures.

Drug residues are excreted primarily in feces and urine in their original form or as one or more active or inactive metabolites, and then released into sewage systems, wastewater and onto the soil, particularly in the case of most veterinary drugs.

The main sources of pharmaceutical water pollution include diffuse sources, primarily from human and animal excrement, which account for the majority of medications released into the environment, and point sources, which result in much more concentrated but geographically limited emissions.

Diffuse sources include improper storage and disposal of expired or unused medications and metabolic excretion via urine or the digestive tract of medications consumed by humans and animals.

Drug substances are grouped into several classes, which themselves encompass a wide range of chemical families as well as degradation metabolites produced by humans or in the environment. These include:

– analgesics and anti-inflammatories, which are the most commonly used molecules in France for human use, such as diclofenac and ibuprofen, acetylsalicylic acid and paracetamol;

– psychotropic drugs, which include antidepressants, anxiolytics and sleeping pills;

– anticonvulsants;

– lipid-lowering drugs of the fibrate family, which are intended to lower the plasma concentration of cholesterol and triglycerides;

– beta-blockers, which lower blood pressure by minimizing the binding of adrenaline to its receptors. They are cardioselective because they act specifically by reducing nerve impulses to the heart;

– antibiotics, which aim to destroy or block the action of bacteria, disrupt, for example, protein synthesis associated with bacterial growth;

– bronchodilators, which act by stimulating beta-2 receptors in the bronchi, causing their short-term dilation.

Other classes that can be mentioned include diuretics, antifungals, antivirals and laxatives.

On the other hand, the use of various antibiotics in animal production and human medicine may explain the observation of distinct correlations between the composition of the bacterial community and antibiotic-resistant populations, and the concentration of tetracycline, sulfonamide, penicillin and fluoroquinolone in wastewater. Indeed, antibiotic-resistant bacteria have been detected in several wastewater sources and clarified wastewater.

Point source contamination includes the direct discharge of pharmaceuticals (and chemicals used in their manufacture) into wastewater from pharmaceutical and chemical industries; the direct or indirect disposal of pharmaceutical agents used in healthcare facilities; the direct dispersal of veterinary drugs into the environment in aquaculture or during the treatment of livestock; indirect dispersal through the spreading of slurry and liquid manure on agricultural land; and the direct dispersal of therapeutic molecules in the form of feed supplements added to fish farms.

These substances are found in different environmental compartments (water, air, soil), through direct or indirect emission (runoff, drainage, atmospheric deposition, etc.), with potential direct or indirect effects on human health and ecosystems, notably via contamination of the food chain (see Figure 1.1). These molecules have a very wide diversity of chemical structures.

When molecules used for therapeutic or diagnostic purposes are not completely degraded in the body, they are excreted primarily in feces and urine in their initial form or as one or more metabolites. These human or veterinary drug residues can, after several stages, be released into the environment. Concentration levels in natural waters vary depending on the chemical stability, biodegradability and physicochemical characteristics of the molecules, as well as the performance of wastewater treatment plants.

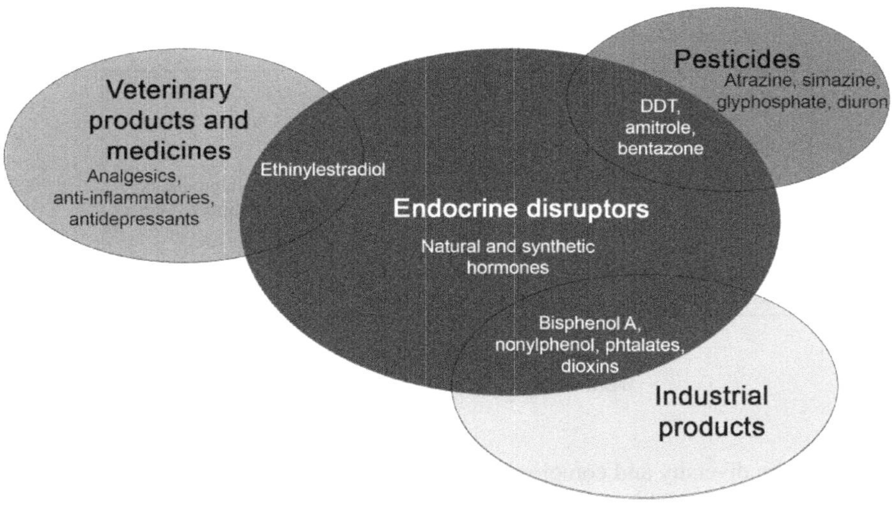

Figure 1.1. *Different categories of micropollutants*

In general, the concentrations of pesticides and pharmaceuticals found in the environment vary from one compound to another, as well as from one site to another. Regional differences in drug use, as well as the biodegradability of pharmaceuticals and their metabolites, could explain these differences in concentrations.

Analysis campaigns conducted at wastewater treatment plant outlets demonstrate the presence of these substances, often in significant concentrations.

	0.001–0.1 µg.L^{-1}	0.1–1 µg.L^{-1}	1–10 µg.L^{-1}	> 10 µg.L^{-1}
> 70%	Terbutaline, diazepam, salbutamol, bromazepam, 17 estradiol, estrone, fluoxetine, nordiazepam	Amitriptyline, metoprolol, roxithromycin, bisoprolol, propranolol, estriol, carbamazepine, diclofenac, gemfibrozil, solalol, sulfamethoxazole Diuron, fluoranthene	Naproxen, atenolol, acebutolol, caffeine, ketoprofen	45 substance s (WFD)
30–70%	Atrazine, simazine, DEA, DIA, tributyltin, pentachlorophenol Indeno(1,2,3-pyrene), endosulfan, isoroturon, benzo(g,h,i)perylene, isodrine, CCl_4	Dichlorophenol, benzothiazole, 4 ter-butylphenol Naphthalene, trichlorobenzene, benzo(b)fluoranthene AMPA, triclosan, tributylphosphate	Di- and trichloromethane, tetrachloroethylene Decabromodiphenyl ether	–
30%	1, dichloroethane, anthracene, benzo(a)pyrene, benzene, dieldrin, hexachlorobenzene Hexabromodiphenyl ether, di- and tribromophenol, tri- and tetrachlorophenol	Benzo(k)fluoranthene, chlorpyrifos, hexachlorocyclohexane, trichloroethylene, pentabromodiphenyl ether Monochlorophenol, glyphosate, tetrabromodiphenylene	C10–13 chloroalkanes, tribromodiphenyl ether	–

Table 1.2. *Example of assessment of water micropollutant concentration*

Due to the diversity and complexity of the substances involved, data on the effects of pharmaceutical residues and endocrine disruptors in the environment, and their prevalence in resources, remain limited. Given the wide variety of molecules considered and the difficulties associated with their analysis, sometimes linked to their low concentrations, studies conducted to date to assess the presence of these compounds in the environment have been limited to a few groups of molecules. These were selected based on production quantities and estimates of their persistence in the environment for pharmaceutical residues and substances (PRS) and were based on potential endocrine activity for endocrine disruptors (EDs), which are molecules that affect the hormonal balance of living species. Recent toxicological studies have demonstrated that, in humans, EDs influence sperm count, genital malformations, impaired sexual function and neuronal development, obesity and even cancer.

Substance	Concentration ($\mu g.L^{-1}$)	Substance	Concentration ($\mu g.L^{-1}$)
Acebutolol	0.3–5.9	Galaxolide	0.1–1.4
Atenolol	0.15–3.2	Gemfibrozil	0.3–1.2
Clofibric acid	0.9	Ibuprofen	0.4–3.5
Bromazepam	0.02–0.12	Iopromide	2–2.9
Carbamazepine	0.1–0.9	Ketoprofen	0.2–3.1
Ciprofloxacin	0.05–1.4	Metoprolol	0.2–1.6
Cyproconazole	0.2–1.9	Naproxen	0.5–4
Diclofenac	0.9–1.2	Nordazepam	0.05–1.1
Estrone	0.05–1.1	Propranolol	0.6–3.8
17-β-estradiol	0.2–1.4	Roxithromycin	0.05–1.5
17-α-estradiol	0.2–1.2	Sulfamethoxazole	0.5–1.3
17-α-ethinylestradiol	0.2–1.2	Sotalol	0.3–1.5
Estriol	0.05–1.1	Terbutaline	0.05–0.1
Erythromycin	0.2–2.9	Triadimefon	0.3–4.2
Fluoxetine	0.3–2.1	Tebuconazol	0.1–3.8

Table 1.3. *Example of micropollutant concentrations in clarified wastewater*

The harmful effects of releases of PE and RSPs into the environment on aquatic and terrestrial organism populations (from simple gastropods to mammals) are now largely undisputed. Numerous examples demonstrate the impact of these substances on fish (incomplete development of genital organs, hermaphroditism, alteration of the male/female ratio, etc.), birds (e.g. thinning of eggshells) and mammals (e.g. decline in seal populations feeding on fish contaminated with PCBs) that are affected. Many questions remain in the scientific community regarding the cause-and-effect relationship, which is often difficult to establish (knowledge is still incomplete regarding the toxic and ecotoxic effects of metabolites, cocktails of substances, mechanisms of disruption, etc.).

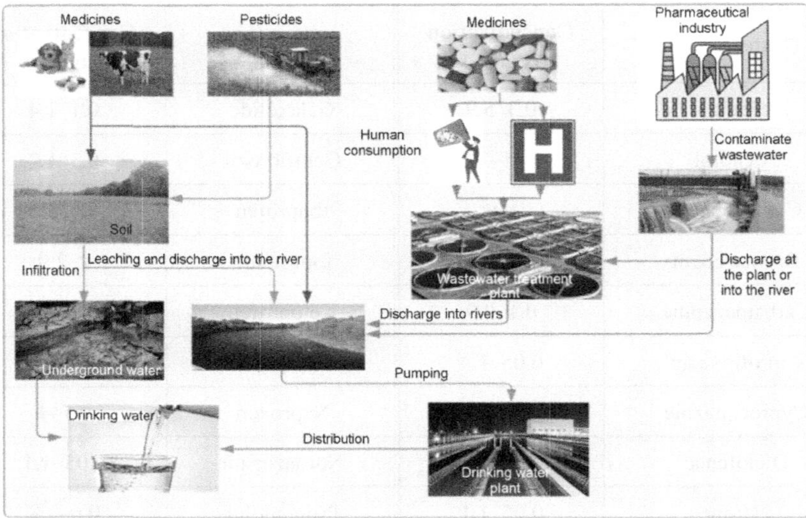

Figure 1.2. *Simplified diagram of the water cycle containing pesticides, drug residues and endocrine disruptors*

According to the European Commission, an endocrine disruptor is an exogenous substance or mixture which alters the functions of the endocrine system and thereby induces adverse health effects in an intact organism, its offspring or subpopulations.

In humans, the links between the presence of pharmaceutical residues and endocrine disruptors in the environment are a cause for concern among the population. European authorities appear to have acknowledged that an association between endocrine disruptors and increasingly common pathologies, such as testicular, breast and prostate cancers, decreased sperm count, malformations of reproductive organs and thyroid disorders, can be confirmed.

Their massive use, as well as the emissions generated during their manufacture or use, result in their presence at concentration levels ranging from a few tens of nanograms per liter to a few micrograms per liter.

The presence of pharmaceutical residues in drinking water is a major concern for the scientific community and public authorities. For several years, the scientific community, public authorities and the public have been questioning the presence in aquatic environments (surface water, groundwater, transfer media, soils) and in drinking water, even in trace amounts, of drug residues and their effects on the environment and human health. This inquiry is part of a broader context of environmental and water resource preservation, particularly against micropollutants, substances that can have a toxic effect on low doses in a given environment (nanograms or micrograms per liter for water).

French public authorities have declared source reduction a priority, even aiming for the complete elimination of emissions of certain substances. Source reduction measures are essential and prevent the spread of substances in the environment, but it will take many years for the effects to be seen, and it will not be possible to control all the risks associated with micropollutants. This is due to the very large number of molecules, uses and practices, as well as to the fact that many micropollutants mainly originate from diffuse pollution emitted by individuals (pharmaceutical residues, hormones, biocides, etc.), which are therefore difficult to eliminate completely.

In Europe, the implementation of a regulatory framework aims to reduce the release of hazardous substances into the natural environment and to help study their fate within these environments. The numerous efforts undertaken, particularly at the European level, contribute to a better understanding of the environmental (chemical) quality of the various available water resources and their evolution.

In France, regulatory requirements aimed at reducing emissions of chemical substances and their presence in the natural environment stem from European regulations and directives, such as the REACH regulation, regulations on biocidal and plant protection products, the Water Framework Directive (WFD) or the Industrial Emissions Directive (IED). These regulations primarily aim to control the placing of potentially hazardous substances on the market, reduce their emissions into various environmental compartments and achieve environmental quality objectives. These increasingly stringent requirements are likely to strengthen in the coming years as scientific knowledge evolves and health and environmental impacts are demonstrated.

Beyond transposing these fundamental European texts into French law, the government has implemented a micropollutant plan to preserve water quality and biodiversity. This plan aims to reduce emissions of micropollutants with proven toxicity, consolidate knowledge and establish lists of pollutants requiring action.

Municipal and industrial wastewater treatment has evolved considerably over the past few decades, discharging effluents according to the increasingly stringent water quality requirements. Unplanned indirect potable water reuse is common worldwide.

However, regarding micropollutants such as emerging contaminants, studies monitoring their presence in surface water or wastewater are becoming increasingly common in the literature.

Wastewater treatment plants, although much improved from a technical aspect, were not designed to eliminate all pharmaceutical molecules. For example, even though paracetamol is broken down by more than 90%, traces can still be found in the clarified wastewater discharged into surface waters. Some compounds, such as diclofenac (an anti-inflammatory) or carbamazepine (an anti-epileptic), are only broken down a little.

These pharmaceutical residues released into surface waters can therefore present a significant environmental risk, as can their metabolites or degradation byproducts, which are sometimes more dangerous than the original product.

1.3. Mechanisms governing the fate of micropollutants in wastewater treatment plants

When transported to wastewater treatment plants, metallic and organic micropollutants are not completely removed from the water by these facilities, which were not designed to eliminate very low pollutant concentrations.

Recent studies show that during treatment, some of these micropollutants are likely to be found in the extracted sludge (Besnault and Martin 2011; Tahar 2013).

Indeed, domestic, industrial, agricultural and medical activities result in the discharge of numerous chemical substances into wastewater treatment plants. Most of these substances are discharged in low concentrations, hence the term "micropollutants" (concentrations of $ng.L^{-1}$ to $\mu g.L^{-1}$).

Micropollutants are thus defined as mineral or organic molecules, originating from agricultural, chemical or industrial processes, which can have negative impacts on ecosystems and the environment beyond a certain threshold.

Wastewater treatment involves transferring organic and mineral pollution from the liquid phase (water) to a solid phase (sludge) and a gaseous phase (emission of CO_2, N_2, etc.). The sludge produced thus constitutes a direct waste product from the treatment process.

The activated sludge treatment process is currently the most common method for treating urban wastewater. Figure 1.4 shows the main stages of an activated sludge treatment process.

Municipal wastewater treatment plants most often include pretreatment, a primary settling stage, biological treatment and a clarifier. Sometimes, tertiary treatment is added when the effluent quality must meet stricter criteria.

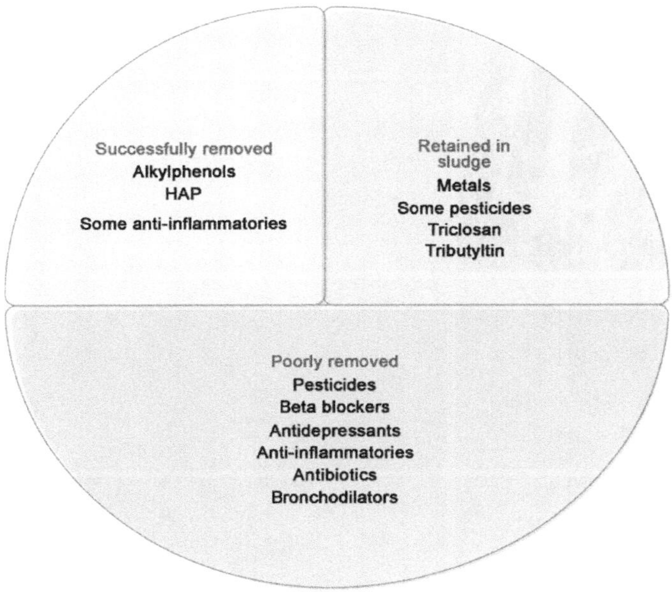

Figure 1.3. *Fate of some substances in wastewater treatment plants*

The biological treatment stage relies on a natural microbiological process, in which microorganisms convert conventional pollutants into end products such as carbon dioxide, water and nitrogen gas and create new bacterial cells.

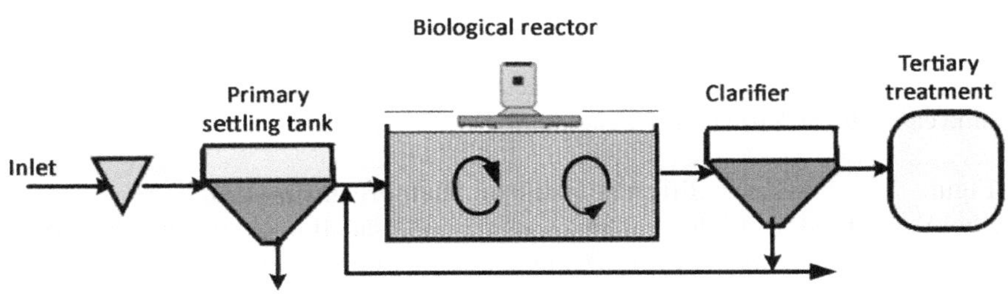

Figure 1.4. *Biological treatment process using activated sludge*

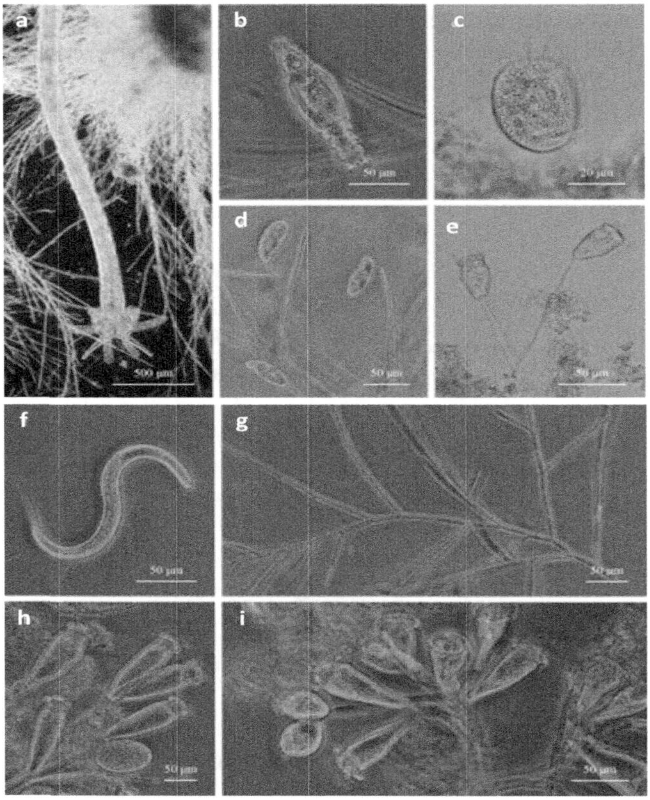

Figure 1.5. *Metazoa (a, b, f), protozoa (c, d, e, h, i) and fungi (g)*

The biological reactor biomass includes bacteria, archaea, fungi, protozoa and metazoa (see Figure 1.5). Protozoa and metazoa are the main predators of bacteria and, along with bacteriophages, contribute to the regulation of the bacterial population (Feng et al. 2017). These organisms have characteristic shapes and are larger than bacteria. The length of bacterial cells is generally approximately 1 μm but can reach, in some cases, up to 5 μm (Eikelboom 2000).

Figure 1.5 shows the different types of microorganisms (fungi, protozoa and metazoa) observed in biological reactors. The degradation and mineralization of nutrients and organic matter occur through a series of enzymatic reactions. Enzymes are intracellular or extracellular proteins that act as biological catalysts. The breakdown of organic matter is carried out by heterotrophic organisms that metabolize organic carbon as a source of energy and growth.

1.3.1. *Molecular characteristics of micropollutants*

Significant variations in the biological elimination of micropollutants (MPs) are reported, ranging from complete removal of some compounds such as ibuprofen to much only slight removal for carbamazepine. The physicochemical properties of MPs help determine the mechanisms of their biological removal. These properties are linked to molecular structure: molar mass, availability of functional groups, branching (Tadkaew et al. 2011). Being chemically complex molecules, the molecular weight, structure, functionality and shape of MPs vary considerably (Rivera-Utrilla et al. 2013). Molecular structure controls the chemical and biochemical reactions that a compound undergoes in a biological reactor. It is generally accepted that the persistence of an organic pollutant is linked to the presence of halogenated elements, nitro-aromatic groups and highly branched molecules in its structure. Conversely, biodegradability increases with compounds containing straight aliphatic chains, esters, acids and hydroxyl functional groups (Howard and Muir 2010). The polarity of molecules also influences the fate of the compound and its hydrophobicity, hence the tendency to favor a hydrophobic compound. The octanol–water parameter is a reference system that provides a commonly recognized measure of hydrophobicity: the logarithm of the partition coefficient, log P (or log Kow). Many MPs include acids, bases, ampholytes and salts.

1.3.2. *Micropollutant removal mechanisms*

The accumulation and transport of micropollutants are governed by various physico-chemical and biological mechanisms (Chipasa 2003):

– biosorption onto cells or extracellular polymers contained in bacterial flocs or biofilms;

– bioaccumulation: this refers to the active interaction between cells and micropollutants;

– sorption onto particulate matter present in wastewater.

Biotransformation induced by biological processes leads to the modification of the molecule and its structure into metabolites or end products (CO_2, H_2O). Intermediate metabolites can be more persistent and less biodegradable than the parent molecule.

While biotransformation alters the structure of the compound, sorption leads to phase transfer. Non-biodegradable substances with a high sorption potential can be discharged with both the clarified effluent and the extracted sludge.

Adsorption onto bacterial flocs results in a longer residence time in the reactor and may lead to subsequent removal via biodegradation in the clarifier or other sludge storage facilities. Other, less significant mechanisms involve the adsorption of particulate matter onto colloidal and inert particles. This phenomenon occurs during pretreatment, primary settling and even clarification stages. Elimination by stripping is rarely observed. This pathway is considered negligible because MPs have a Henry coefficient (H) < 10^{-5}.

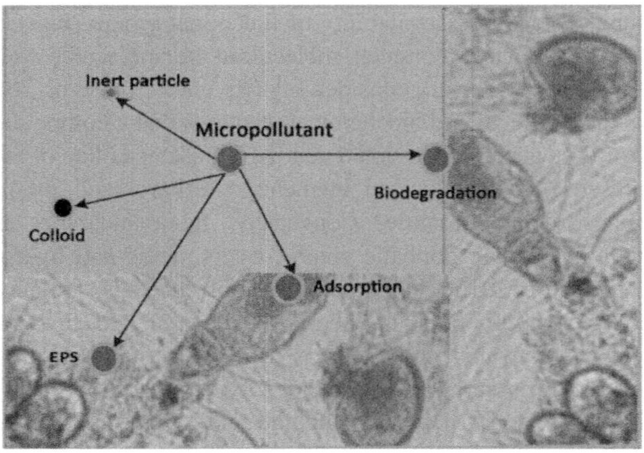

Figure 1.6. *Micropollutant removal types during biological treatment*

1.3.2.1. *Biotransformation*

Microbial biotransformation refers to the microbial conversion of a compound into a transformed product called a "metabolite" or "metabolic intermediate". Different intermediates can be formed either by the degradation of the parent substance or by its modification. These metabolic reactions can be intracellular or extracellular (Fischer and Majewsky 2014; Stadlmair et al. 2018). Biotransformation is considered an important pathway for the neutralization of microplastics, even though it is still poorly understood and the compounds can be completely mineralized or transformed into metabolites, which could be even more problematic for the environment than the parent molecule, especially when the biologically active part remains intact during degradation (Boxall et al. 2004; Magdeburg et al. 2014).

MPs are generally recognized as non-growth substrates because their concentrations are too low to participate in bacterial metabolism. Co-metabolous biodegradation is thought to be the dominant factor in the biodegradation process (Nsenga Kumwimba 2019), This means that the MPs are biodegraded by non-specific enzymes generated by the metabolism of the main substrates in wastewater (Fischer and Majewsky 2014).

1.3.2.2. *Biosorption on biomass*

Biosorption is a process that involves the retention of organic substances on solid materials of biological origin (Yuncu et al. 2006; Pagnanelli et al. 2009).

Various mechanisms are involved, such as surface complexation where carboxyl, hydroxyl and phosphoric functional groups form binding sites on bacterial surfaces, biosorption onto cell walls and external surfaces, and then active transport across the cell membrane.

Biosorption can be defined as a metabolically passive process, physico-chemical in nature, involving ionic or molecular bonds. The attraction of micropollutant molecules towards the solid biomass matrix is governed by intramolecular interactions between the compound and the biomass surface. They are hydrophobic and/or electrostatic, also involving mechanisms such as van der Waals interactions, hydrogen bonding, electron donors, electrostatic interactions of cations and anions with charged sites in biomass, cation exchange, cation bridging and ligand exchange with surface hydroxyl groups (OH^-) (Trapp et al. 2010; Hyland et al. 2012; MacKay and Vasudevan 2012; Oberoi et al. 2019; Burzio 2022).

Hydrophobicity and the ionic state of MPs help to determine the most predominant sorption mechanisms (Rybacka and Andersson 2016). The characteristics of the liquid medium, in terms of pH and ionic strength, influence the surface chemistry of the adsorbent and the activity of the compound, and therefore play a role in sorption dynamics. Since these interactions involve several functional groups, the characteristics of the biomass in the biological reactor (surface loading, specific surface area, EPS content, mineral content, degree of oxidation of the organic matter) determine the degree of adsorption (Barret 2010a; Alvarino et al. 2018a, 2018b; Zhang et al. 2018).

Since biomass is generally considered to be negatively charged at the surface (Hermansson 1999), and since MPs are positively charged, micropollutants are likely to exhibit a high sorption potential due to electrical attraction (Franco and Trapp 2008). The lipophilic cell membrane, the cell wall of microorganisms and the lipid fraction of sludge attract hydrophobic compounds (Verlicchi et al. 2012a, 2012b). Different sorption capacities are therefore observed with different types of biomasses (primary and secondary sludge, digested sludge, biofilms, flocs, anaerobic sludge, etc.).

The distribution of MP between the aqueous phase and the biomass is generally assumed to be instantaneous (Pomiès 2013) and involves two opposing mechanisms: sorption from the aqueous phase to the solid phase and desorption from the solid phase to the aqueous phase (Joss et al. 2006). Equilibrium is reached when the sorption kinetics are identical to those of desorption.

The hydraulic residence time in the biological reactor affects the contact time between the biomass and the MPs. Considering the slow kinetics of MP removal, a long residence time is beneficial and positively influences the degradation of the compounds (Gros et al. 2010).

Similarly, increasing the age of sludge has a beneficial impact on the removal of particulate matter. Microbial diversity increases the number of metabolic pathways (Falås et al. 2016) and promotes the growth of slow-growing microorganisms such as nitrifiers, which contribute to the degradation of a wide range of MPs (Park et al. 2017).

Temperature affects microbial activity and the sorption of several MPs in biomass. (Xu et al. 2008; Zeng et al. 2009; Tran et al. 2013). The pH of the environment determines the protonation state of the compounds, affecting their solubility and sorption capacity on biomass (Trapp et al. 2010).

The biomass configuration, whether in free or fixed culture, can differentiate the performance of the biological reactor. Higher MP removal rates have been observed in fixed culture biofilters (MBBRs) compared to activated sludge (Falås et al. 2012a, 2012b). Longer biomass retention in the biofilm favors slow-growing organisms, as does the presence of microniches with redox conditions that stimulate the development of specific microbial communities (Wolff et al. 2021). In a biofilm, microorganisms integrated into an EPS matrix, which is more developed than in activated sludge, facilitate the bio-absorption of MPs (Schmidt et al. 2012). EPSs, being a complex mixture of polymers, offers rich bonding sites due to the presence of anionic, cationic and nonpolar groups (Wunder et al. 2011; Zhang et al. 2018). EPSs have been observed to be key components in the sorption elimination of OMPs (Xu et al. 2013; Zhang et al. 2018).

1.3.2.3. *Biokinetics of micropollutant elimination*

Various studies show that in wastewater treatment plants, certain organic micropollutants can be directly degraded by specific bacteria. The overall mass of organic matter is eliminated by the diverse biomass present in the biofilm or activated sludge.

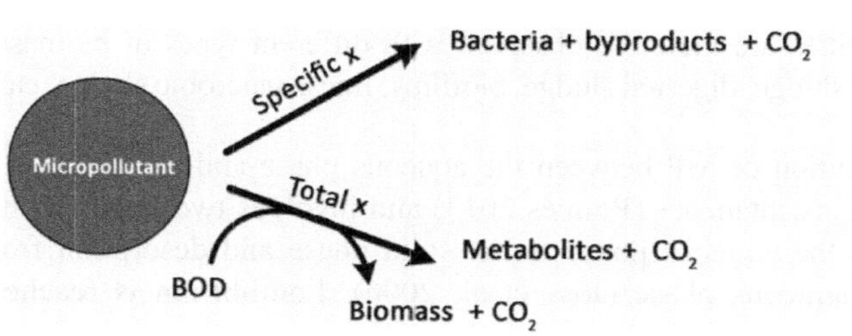

Figure 1.7. *Schematic diagram of the degradation of micropollutants and BOD*

The fate of micropollutants in wastewater treatment plants is governed by various processes, including solubilization (characterized by log Kow, water/octanol partition coefficient, for organic trace compounds), sorption/desorption to the biological matrix (characterized by the Kd coefficient), degradation for organic micropollutants (characterized by the kbiol) and finally, for some compounds, volatilization (stripping) characterized by the Henry's constant (H) of the molecule.

1.3.2.3.1. Sorption constant (Kd)

The mechanisms of sorption of organic compounds onto solid matrices (wastewater) involve two processes: adsorption of the compound from the liquid phase onto the surface of the solid and distribution of the compounds between the aqueous phase and the organic matter contained in the solid phase.

The equilibrium exchanges between the aqueous and solid phases of the adsorbent are represented by the sorption distribution coefficient $Kd_{solid,}$ defined according to the following equation (Tolls 2001):

$$Kd_{solid} = Cs/Caq$$

where Cs is the concentration of the adsorbed compound ($\mu g.g^{-1}$) and Caq is the concentration of the compound in the aqueous phase ($\mu g.L^{-1}$).

Regarding the sorption phenomenon, the MP (micropollutant) concentration sorbed on the biomass (C_{bio}, $\mu g.L^{-1}$) is proportional to the dissolved concentration (C, $\mu g.L^{-1}$).

The distribution coefficient between biomass and effluent (Kd) is expressed in the form (Torresi et al. 2017):

$$\mathrm{Kd} = \frac{\mathrm{Cbio}}{\mathrm{C\,X}} = \frac{\mathrm{Co} - \mathrm{Cf}}{\mathrm{Cf\,X}}$$

The concentration absorbed by the biomass (Cbio, $\mu g.L^{-1}$) is estimated from the MP concentration at the treatment plant inlet (Co, $\mu g.L^{-1}$) and the final concentration (Cf, $\mu g.L^{-1}$). This linear equation is a simplified case of the empirical Freundlich isotherm (Schwarzenbach et al. 2003).

In theory, the sorption of substances onto the sludge with $\log Kd < 2$ is negligible, whereas the sorption of substances onto the sludge with log Kd values > 4 is one of the main removal processes (Clara et al. 2004; Joss et al. 2005). Clara et al. (2004) and Joss et al. (2005) claimed that for compounds whose sorption coefficient (Kd) is below 300 L/kg, adsorption onto the sludge is not significant, and the transformation can be easily assessed by a balance calculation comparing the concentrations at the inlet and outlet of the process.

Sorption models have been established, proposing an approach that takes into account the role of dissolved and colloidal matter (Barret 2010a). A micropollutant present in the bacterial floc can be distributed according to three possible phases (see Figure 1.8): the dissolved free phase (concentration C_{free} in $\mu g.L^{-1}$), the phase sorbed to the dissolved and colloidal matter (concentration C_{col} in $\mu g.g^{-1}$) and the particulate phase (concentration C_{part} in $\mu g.g^{-1}$).

Thus, at equilibrium, the system can be described by the equations:

$$K_{part} = C_{part}/C_{free}$$

$$K_{col} = C_{col}/C_{free}$$

K_{part} is the sorption equilibrium constant of the micropollutant to particles (in $L.g^{-1}$) and K_{col} is that of sorption to colloidal matter (in $L.g^{-1}$).

The concentration C_{free} is difficult to obtain experimentally, unlike concentrations C_{part} and C_{aq} (concentration of dissolved micropollutants, which corresponds to the addition of the dissolved and sorbed phases to the colloidal matter, in $\mu g.L^{-1}$).

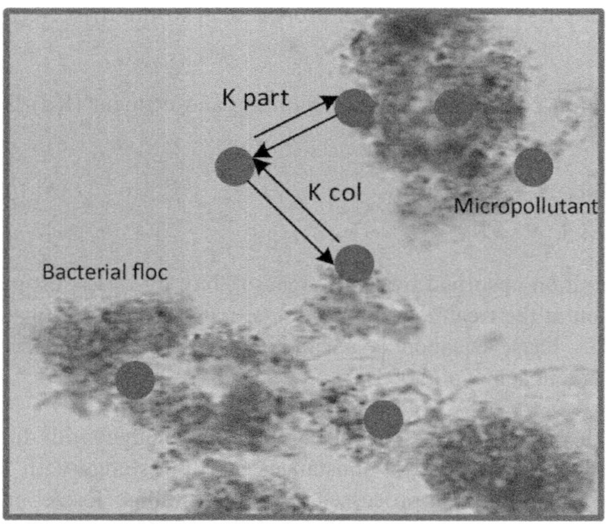

Figure 1.8. *Representation of the adsorption of a micropollutant by a bacterial floc (Barret 2010a)*

After determining C_{aq} and C_{part}, the global partition coefficient K can be calculated, and then K_{part} and K_{col} can be estimated:

$$\text{Kpart} = \frac{\text{Cpart}}{\text{Caq}} = \frac{\text{Kpart Cfree}}{\text{Kcol Cfree (Ccol)} + \text{Cfree}} = \frac{\text{Kpart}}{\text{Kcol (Ccol)} + 1}$$

where C_{col} is the concentration of the dissolved and colloidal matter ($g.L^{-1}$).

1.3.2.3.2. Biodegradation constant (Kbiol)

The degradation or transformation rates of micropollutants are calculated using the following equation, assuming pseudo-first-order kinetics (Joss et al. 2006):

$$dC/dt = kbio\ X\ C$$

where C is the concentration of the compound ($\mu g.L^{-1}$), t is the time (d), kbio is the rate constant of degradation or transformation ($L\ g^{-1}\ SS.d^{-1}$) and X is the concentration of suspended matter ($g\ SS.L^{-1}$). The kbio values include sorption, desorption and biological transformation.

Three groups of micropollutants can be identified according to their biodegradation constant kbiol (Joss et al. 2006):

– Compounds with kbiol < 0.1 L/g/d are not eliminated in significant quantities (< 20%).

– More than 90% of compounds with kbiol > 10 L/g/d are transformed.

– For compounds with 0.1 L/g/d < kbiol < 10 L/g/d, a moderate transformation is expected.

On the other hand, the biodegradation of organic micropollutants is often described by a first-order law (Pomiès 2013):

$$(dLmp/dt)\ biodegradation = -\ kbmp\ Lmp\ (t)\ Xa$$

or:

$$(dLmp/dt)\ biodegradation = -\ k'bmp\ Lmp\ (t)$$

where Lmp is the concentration of micropollutants in the dissolved phase ($M.L^{-3}$), kbmp is the biodegradation rate constant of the micropollutant ($L^3.M^{-1}.T^{-1}$), k'bmp is the kinetic biodegradation constant of the micropollutant (T^{-1}) and X is the concentration of active biomass ($M.L^{-3}$). Most of the methods described for determining biodegradation are based on measuring the disappearance of the compound over time in batch mode, after boosting the biomass and adding wastewater, either under aerobic or anoxic conditions to reproduce the operating conditions of a wastewater treatment plant (Pomiès 2013; Seira 2013). The kinetic constants of biodegradation are then determined from the slopes of change in micropollutant concentrations.

1.3.2.3.3. Octanol–water partition coefficient (Kow)

During the primary and secondary settling stages, hydrophobic compounds can adsorb onto suspended solids and bacterial flocs via a mechanism linked to van der Waals forces and interactions between the micropollutant and the bacterial surface.

Solubility and hydrophobicity allow us to estimate the feasibility of molecule adsorption. The adsorption of micropollutants is primarily governed by hydrophobicity and can therefore be predicted using the log Kow.

The octanol/water partition coefficient (Kow) is defined as the ratio of the concentration of a chemical in the octanol phase to its concentration in the aqueous phase of a two-phase octanol/water system. Log Kow is equal to the logarithm of the ratio of the concentrations of the substance in octanol to those in water. This parameter allows us to understand the hydrophilic or hydrophobic nature of a molecule. Indeed, if the log Kow is positive and very high, this indicates that the molecule in question is much more soluble in octanol than in water, reflecting its hydrophobic nature. Conversely, if the log Kow is negative, it means that the molecule in question is hydrophilic. Kow values are dimensionless. In terms of log Kow, the range varies between –3 and 7. The log Kow is generally inversely related to water solubility and directly proportional to the molecular weight of a substance. Due to its increasing use in estimating the adsorption of organic molecules on activated carbon, Kow is considered a necessary parameter.

Log D is a more refined measure that takes into account its change depending on the pH of the aqueous phase and the pKa of the compound being studied. For neutral (uncharged) compounds, log D and log Kow are equal. A positively or negatively charged molecule has a log D value lower than log Kow.

Table 1.4 presents log Kow values for pharmaceuticals, endocrine disruptors, industrial products, pesticides, disinfection byproducts, etc. Substances with a log Kow greater than 3 will be very well adsorbed. Those with a log Kow between 1 and 3 will be fairly well adsorbed. Molecules with a log Kow between 0 and 0.5 will be poorly adsorbed, and those with a negative log Kow will not be adsorbed at all.

The structure of the molecule influences adsorption kinetics. When a molecule is not symmetrical, adsorption is less efficient than with a symmetrical molecule. The

OH, NH_2, SO_2 and COOH functional groups tend to reduce the adsorption mechanism as well as the ionized molecules. On the other hand, the functional groups NO_2, NOH, COOR and X, molecules with double bonds or delocalized electrons facilitate adsorption mechanisms.

The solubility of organic micropollutants varies over a wide range of 0.15 $mg.L^{-1}$ (antidepressant maprotiline) to 270,000 $mg.L^{-1}$ (artificial sweetener acesulfame).

Several correlations have been observed between the log Kow and the sorption of organic compounds onto bacterial flocs. The compounds can be classified according to their adsorption capacity based on their Kow:

– log Kow < 2.5: low sorption potential;

– 2.5 < log Kow < 4.0: average sorption potential;

– log Kow > 4.0: high sorption potential.

Compounds and log Kow		
Acenaphthene 3.90	Diazinon 3.01	Metolachlor 3.25
Acetochlor 3.17	Dibromochloromethane 2.16	Metolachlor EZA –0.03
Acetochlor EZA 1.82	Dibutyltin cation 4.44	Metolachlor OSA –0.03
Acetochlor OSA 1.82	Dicamba –1.88	Metamitron 0.85
Aclonifen 3.09	Diclofenac 0.70	Metazachlor 2.50
Alachlor 2.64	1,2-Dichlorobenzene 3.38	Metsulfuron –1.70
Aldicarb 1.13	1,4-Dichlorobenzene 3.38	MIB 3.31
Aldrin 3.01	1,2-Dichloroethane 1.45	Methylparathion 3.32
Ametryn 2.98	Dichloromethane 1.25	Methoxychlor 5.08
Aminotriazole –0.97	Dichlorprop 2.29	2,4-MCPA –0.81
AMPA –2.17	Dichlorvos 1.90	Monuron 1.46

Compounds and log Kow		
Anthracene 4.54	Dicofol 4.30	Naphthalene 3.36
Atrazine 2.33	Dieldrin 5.40	Napropamide 3.36
Azithromycin 4.02	Diflufenican 4.20	Nicosulfuron 1.00
Azoxystrobin 2.50	Dimethenamid 2.47	Nonylphenol 5.00
Benazolin 1.34	Dimethachlor 2.17	Octylphenols 5.30
Bentazone –0.46	Dinoseb 3.15	Oxadiazon 5.30
Benzene 2.11	Diquat –4.60	Parathion 3.81
Benzo (a) pyrene 6.35	Diuron 2.81	Paracetamol 0.46
Benzo (b) fluoranthene 6.67	Endosulfan 3.62	PCB-24 5.10
Benzo (k) fluoranthene 6.30	Endrin 3.20	PCB-44 5.58
Benzo (g,h,j) perylene 6.61	Erythromycin 4.51	PCB 118 4.4–5.0
Benzothiazol 2.10	Esteron 3.13	Pendimethalin 5.20
Bezafibrate 4.25	Estriol E3 2.45	Pentachlorobenzene 5.20
BDE 047 6.77	Ethephon –1.89	Pentachlorophenol 3.00
Bifenox 3.64	Ethofumesate 2.70	PEFOS 3.5–5.8
Biphenyl 4.09	Epoxiconazole 3.30	Phenanthrene 4.57
Boscalid 2.96	17-alpha-ethinylestradiol 3.67	Phenmedipham 3.59
Bromacil 2.02	17-beta-estradiol 4.01	Prochloraz 4.30

Compounds and log Kow		
Bromobenzene 2.99	Fenuron 1.00	Prometon 2.99
Monobutyltin cation 3.31	Fenoprop 3.80	Propamocarb HCl –1.20
Carbamazepine 2.45	Flufenacet 3.20	Propyzamide 3.00
Carbendazim 1.52	Fluorobenzene 2.27	Propazine 2.94
Carbaryl 2.81	Fluoranthene 5.10	Prosulfocarb 4.48
Carbofuran 2.07	Fluorene 4.20	Prothioconazole 3.82
Clarithromycin 3.16	Fluoxetine 4.05	Pyrene 5.18
Chlordane 3.32	Gemfibrozil 4.77	Quinoxyfen 4.70
Chloroalkanes (C10-13) 7.00	Genistein 2.84	Quinmerac 1.76
Chlorobenzene 2.84	Geosmin 3.10	Simazine 2.16
Chloroform 1.97	Glyphosate –3.20	Spiroxamine 2.80
Chlormequat chloride –1.56	Heptachlor 4.41	Sulcotrione –1.70
Chlorotoluron 2.41	Heptachlor epoxide 2.70	Sulfamethazine 3.11
Chlorpyrifos 3.31	Heptachlorobenzene 5.73	Sulfamethoxazole 0.89
Chloprophamn 3.06	Hexachlorobutadiene 4.78	Tebuconazole 3.70
Chlorsulfuron –0.99	YHCH lindane 3.72	Terbacil 1.89
Chlorothalonil 2.94	Hydroxyalachlor 2.74	Terbutryn 3.65
Clofenvinphos 4.22	Hydroxymetholachlor 2.81	1,2,4-Trichlorobenzene 4.02

Compounds and log Kow					
Chlorpyriphos	4.23	Ibuprofen	3.97	Trietazine	3.35
Ciprofloxacin	0.28	Indeno (1,2,3-cd) pyrene	7.60	1,1,2,2-Tetrachloroethane	2.39
Cyanazine	2.24	Iodobenzene	3.25	1,1,1-Trichloroethane	2.47
Cybutryne	2.70-4.00	Iodosulfuron	1.59	Triclosan	4.76
Cypermethrin	5.50	Isodrine	6.50	Tetrachloroethylene	2.53
Cyprodinil	4.00	Ioxynil	2.20	Carbon Tetrachloride	2.60
DDT	6.90	Imazapyr	0.11	Thiram	1.73
2.4. D	2.81	Imidaclopride	0.57	Tributylphosphate	4.00
Deethylatrazine	1.51	Isoproturon	2.50	Triclopyr	–0.45
DEHP	7.60	Linuron	2.19	Trichlorobenzene	4.05
Deisopropylatrazine	1.15	Malathion	2.89	Trichloroethylene	2.99
Deschloroalachlor	3.08	Mancozeb	1.30	Trifuran	5.07
Deschlorometholachlor	3.10	Mecoprop	–0.19	Trifluralin	5.34
Desethylhydroxyatrazine	–2.70	Mesotrione	0.11	Tylosin	0.14
		Metaldehyde	0.12		

Table 1.4. *Partition coefficient values Kow for various organic micropollutants*

Name	Active ingredient	Effluent ($ng.L^{-1}$)	Solubility ($mg.mL^{-1}$)	pKa	Henry's constant (atm. m^{-3}.mole)
Atenolol	Beta blocker	843	0.3	9.6	1.37 E-18
Azithromycin	Antibiotic	175	< 1–25°C	8.74	5.30 E-29
Bezafibrate	Hypolipidemic	139	0.00155	3.83	–
Carbamazepine	Anticonvulsant	482	0.152	15.96	1.08×10^{-10}
Carbamazepine-10, 11-dihydro-10, 11-dihydroxy	Transformation product	1,551	–	–	–
Clarithromycin	Antibiotic	276	0.00033	8.99–25°C	1.73 E-29
Diatrizoate (amidotrizoic acid)	Contrast agent	598	0.107	2.17	–
Diclofenac	Analgesic	647	0.00447	4	4.73 E-12
Erythromycin	Antibiotic	42	0.459	12.44	1.46 E-29
Ethinylestradiol	Estrogen	2	0.00677	10.33	7.94 E-12
Ibuprofen	Analgesic	394	0.0684	4.85	1.50 E-07
Iomeprol	Contrast agent	380	–	–	–
Iopamidol	Contrast agent	377	0.117	4.15	1.14 E-25
Iopromide	Contrast agent	876	0.0238	–	1.00 E-28
Mefenamic acid	Analgesic	870	0.0137	3.89	2.57 E-11
Metformin	Antidiabetic	10,347	1.38	12.4	–
Metoprolol	Beta blocker	166	0.402	14.09	1.40 E-13
Naproxen	Analgesic	462	0.0511	4.19	3.39 E-10
Sotalol	Beta blocker	435	0.782	10.07	2.49 E-14
Sulfamethoxazole	Antibiotic	238	0.459	6.16	6.42 E-13
N4-acetyl-sulfame-thoxazole	Transformation product	67	–	–	–
Trimethoprim	Antibiotic	100	0.615	17.33	2.39 E-14
Penicillin V	Personal care	28.7	0.454	3.39	4.42 E-15
Irbesartan	Antihypertensive	479.5	0.00884	7.4	–
Tramadol	Analgesic	255.8	0.75	13.8	1.54 E-11
Risperidone	Neuroleptic	6.9	0.171	8.76	–
Trihexyphenidyl	Anti-dementia	0.2	0.00314	13.84	4.73 E-10

Name	Active ingredient	Effluent ($ng.L^{-1}$)	Solubility ($mg.mL^{-1}$)	pKa	Henry's constant (atm. m^{-3}.mole)
Venlafaxine	Antidepressant	118.9	0.23	14.42	–
Codeine	Morphine derivative	70.6	0.577	13.78	7.58 E-14
Fluconazole	Antifungal	108.2	1.39	12.71	–
Diphenhydramine	Antihistamine	11.7	0.0752	8.98	3,70 E-09
Repaglinide	Antidiabetic	3.1	0.00294	3.68	–
Flecainide	Antiarrhythmic	45.5	0.0324	13.68	5.75 E-13
Bisoprolol	Beta blocker	41.6	0.0707	14.09	2.89 E-15
Alfuzosin	Alpha-blocker	2.8	0.282	14.64	–
Bupropion	Antidepressant	1.0	312	18.29	–
Ciprofloxacin	Antibiotic	96.3	1.35	6.09	5.09 E-19
Oxazepam	Anxiolytic	161.7	0.0881	10.61	5.53 E-10
Carbamazepine	Antiepileptic	832.3	0.152	15.96	1.08 E-10
Diclofenac	Analgesic	647	0.00447	4	4.73 E-12
Orphenadrine	Antihistamine	3.9	0.03	8.91	4.08 E-09
Sulfamethoxazole (VITO)	Antibiotic	280.2	0.459	6.16	–
Haloperidol	Psychiatric medication	32.2	0.00446	8.66	2.26 E-14
Citalopram	Antidepressant	3.8	–	–	–
Sulfamethoxazole (JRC)	Antibiotic	142.3	0.459	6.16	–
Fexofenadine	Antihistamine	165	0.00266	4.04	–
Diltiazem	Antiarrhythmic	10.7	0.0168	12.86	8.61 E-17
Fluoxetine	Antidepressant	2.1	0.0017	9.8	8.90 E-08
Terbutaline	Anti-asthmatic	1.1	5.84	8.86	1.65 E-18
Clindamycin	Antibiotic	70.4	3.1	12.16	2.89 E-22
Telmisartan	Antihypertensive	367.5	0.0035	3.65	–
Eprosartan	Antihypertensive	226.8	0.00866	3.63	–
Gemfibrozil	Hypolipidemic	137.7	0.0278	4.42	–
Zolpidem	Hypnotic	1.5	0.0313	6.2	–
Hydroxyzine	Antihistamine	1.1	0.0914	15.12	–

Name	Active ingredient	Effluent (ng.L^{-1})	Solubility (mg.mL^{-1})	pKa	Henry's constant (atm. m^{-3}.mole)
Ketoprofen	Analgesic	86.0	0.0213	4.45	2.12 E-11
Ranitidine	Antihistamine	68	0.0795	8.08	3.42 E-15
Triclosan	Disinfectant	74.8	0.00605	7.9	4.99 E-09
Levamisole	Anthelmintic	40.6	1.44	6.98	4.03 E-10
Lincomycin	Antibiotic	31.2	29.3	12.37	3.00 E-23
Rosuvastatin	Statin	31.0	0.0886	4	–
Mianserin	Antidepressant	1.5	0.232	6.92	–
Clofibric acid	Hypolipidemic	5.3	0.583	– 4.9	2.19 E-08
Iohexol	Radioactive contrast agent	158	0.796	11.73	2.66 E-29
Memantine	Anti-dementia	22.8	0.0455	10.7	1.47 E-05
Sertraline	Antidepressant	2.1	0.000145	9.85	–
Tiamulin	Antibiotic	3.3	–	–	–
Clonazepam	Anticonvulsant	1.6	0.0106	11.89	7.02 E-13
Alprazolam	Antidepressant	1.3	0.0324	18.3	9.77 E-12
Fenofibrate	Hypolipidemic	1.1	0.000707	– 4.9	–
Sulfadiazine	Antibiotic	3.5	0.601	6.36	1.58 E-10
Tilmicosin	Antibiotic	3.1	–	–	–
Cyproheptadine	Chemotherapy	3.9	0.0136	8.05	9.20 E-09
Methylbenzotriazole	Personal care	2900	0.366	8.55	4.13 E-07
Gadolinium	Personal care	115.0	–	–	–
Loperamide	Personal care	29.3	0.00086	13.96	–
Buprenorphine	Personal care	3.9	0.0168	8.31–25°C	1.76 E-17
Maprotiline	Personal care	0.4	0.00015	10.54	–
Duloxetine	Personal care	0.1	0.00296	9.7	–
Miconazole	Personal care	0.2	0.000763	6.77	–
Chlorpromazine	Personal care	0.1	0.00417	9.3–25°C	3.95 E-11
Flutamide	Personal care	0.1	0.00566	13.17	3.73 E-10
DEET, N,N'-diethyl-toluamide	Personal care	678.1	0.912	–	2.08 E-08

Name	Active ingredient	Effluent ($ng.L^{-1}$)	Solubility ($mg.mL^{-1}$)	pKa	Henry's constant (atm. m^{-3}.mole)
Caffeine	Food additive	191.1	11.0	10.4–40°C	1.90 E-19
Acesulfame	Food additive	22,500	588	5.67	–
Sucralose	Food additive	4,600	22.7	4.2	–
Diazinon	Insecticide	173	0.04	2.6	1.13 E-07
Diethyl-toluamide (DEET)	Insecticide	593	0.912	–	2.08 E-08
Dimethoate	Insecticide	–	25	–	1.05 E-10
MCPA	Insecticide	149.9	0.63	3.13	1.33 E-09
Carbaryl	Insecticide	1.6	0.11	10.4	–
2,4-D	Herbicide	13	0.012	2.73	1.59 E-07
Carbendazim	Fungicide	81	0.029	4.2	2.12 E-11
Diuron	Herbicide	201	0.042	–	5.04 E-10
Glyphosate	Herbicide	–	12	0.8	4.08 E-19
Irgarol (cybutryne)	Herbicide	30	–	–	–
Isoproturon	Herbicide	12	0.065	–	1.12 E-10
MCPA	Herbicide	25	0.63	3.13	1.33 E-09
Mecoprop-P	Herbicide	424	0.62	3.1	1.82 E-08
Triclosan	Microbicide	116	0.010	7.9	4.99 E-09
Terbuthylazine	Herbicide	90.6	0.0085	2	3.72 E-08
Atrazine	Herbicide	4.2	0.0347	1.7	2.36 E-09
Terbutylazine-desethyl	Herbicide	68.8	–	–	–
Isoproturon	Herbicide	10.1	0.065	–	1.12 E-10
Bentazone	Herbicide	9.6	0.5	2.92	2.18 E-09
Metolachlor	Herbicide	12.4	0.53	–	9 E-09
Dichlorprop	Herbicide	9.6	0.35	3.1	8.68 E-11
Simazine	Herbicide	26.3	0.0062	1.62	9.42 E-10
Atrazine-diethyl	Herbicide	13.8	3.2	–	1.53 E-09
Chlortoluron	Herbicide	3.2	0.07	–	–
Hexazinone	Herbicide	0.8	33	–	2.26 E-12
Linuron	Herbicide	40.1	0.075	–	–

Name	Active ingredient	Effluent (ng.L^{-1})	Solubility (mg.mL^{-1})	pKa	Henry's constant (atm. m^{-3}.mole)
2,4,5-T	Herbicide	0.3	0.248	2.88	6.83 E-09
Bisphenol A (BPA)	Additive	331	0.12	9.6	1 E-11
Estradiol	Estrogen	3	0.0213	10.33	3.64 E-11
Estrone	Estrogen	15	0.00394	10.33	3.8 E-10
Nonylphenol	Additive	267	0.00635	10.25	1.1 E-06
Perfluorooctane sulfonate (PFOS)	Surfactant	–	3.1	0.14	–

Table 1.5. *Concentration of organic micropollutants at the outlet of wastewater treatment plants (activated sludge)*

1.4. Effectiveness of biological treatments

The biological activated sludge process with extended aeration shows that approximately 50% of substances are removed, with highly variable efficiencies ranging from 20% to 95%. Some substances are primarily removed by biodegradation (triclosan, estrogenic hormones, analgesics), but most are removed by adsorption (particularly PBDEs, PAHs, or certain metals such as Hg, Fe, Cu, Cr, Zn, Cd, Pb).

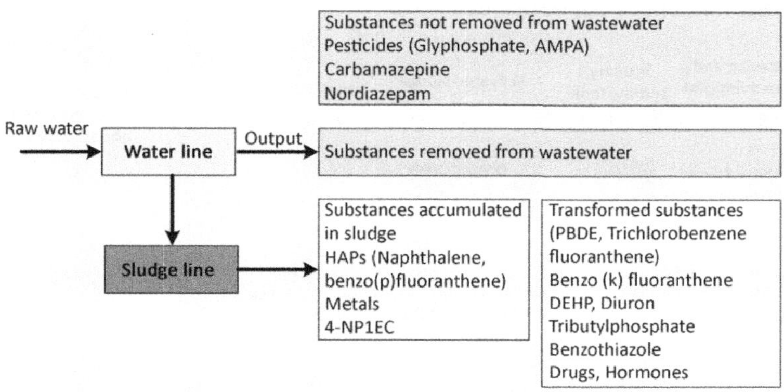

Figure 1.9. *Fate of micropollutants in a conventional biological wastewater treatment process (Besnault et al. 2009)*

Some substances are eliminated at a rate of less than 20%, meaning that biological treatment has proven ineffective or only marginally effective. These include polar pesticides (glyphosate, AMPA (a glyphosate degradation product), diuron), some pharmaceutical substances (carbamazepine, diclofenac, propranolol) and alkylphenol carboxylates, the latter showing negative yields because they are produced by the biological oxidation of alkylphenols.

In total, 15% of priority substances, 30% of organic molecules and 90% of pharmaceutical substances are found in conventional wastewater treatment plant discharge in concentrations exceeding 100 $ng.L^{-1}$ due to their physicochemical properties and their high concentration at the inlet of the wastewater treatment plants (Besnault and Martin 2011).

1.4.1. *Organic micropollutants*

An activated sludge floc is characterized by a microstructure, resulting from bioflocculation by microorganisms (small and mechanically fragile flocs), and a macrostructure, where filamentous microorganisms create a three-dimensional network, a support to which the small flocs can adhere, thus forming larger and stronger flocs (see Figure 1.10).

The cells and extracellular matter that make up the sludge clump together and adopt a structure resembling a multi-dimensional fractal geometry (Chu 2004). An initial organization into flocs approximately 50–100 μm in diameter, 60% porosity and a pore diameter of approximately 10 μm can be observed, and at a lower level, much denser microflocs can be distinguished.

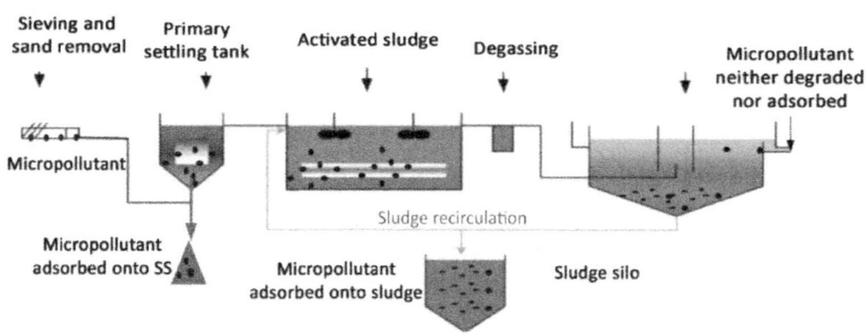

Figure 1.10. *Fate of micropollutants in a wastewater treatment plant*

Adsorption onto suspended solids in primary and secondary treatment units is an important process for PM removal. Adsorption can occur due to hydrophobic interactions between the aliphatic and aromatic groups of compounds with the fatty and lipid fractions in primary sludge and the lipophilic cell membrane of microorganisms in secondary sludge, respectively. Electrostatic interactions also occur between positively charged micropollutant groups and negatively charged microorganisms in the secondary sludge. Many acidic pharmaceuticals are negatively charged at neutral pH, and their sorption onto the sludge is negligible.

With a nonpolar core and a polar fraction, the properties of pharmaceuticals and antibiotics vary considerably, making it difficult to estimate their sorption onto sludge. Nine different biosolids are produced by municipal wastewater treatment plants in seven US states. Sludge is a complex biological matrix containing numerous microorganisms, bacteria (10^{12}–10^{13} bacteria/g) primarily, as well as protozoa and fungi. This living portion represents only 5–20% of the total matter, the remaining 80–95% being made up of extracellular organic and mineral matter (Nielsen et al. 2004).

The adsorption of micropollutants onto the sludge did not show any particular trend, and weak correlations were found between the concentrations of organic micropollutants in the sludge and the log Kow value. This suggests that the organic carbon content of the biosolids may not be the only factor controlling micropollutant adsorption. Compounds with low water solubility and a low Kow value are generally expected to be more likely to be present in sludge than highly soluble micropollutants. The biological removal of micropollutants is complex. Several factors must be considered: the amount of organic matter entering the influencing stream (which generally varies over time and space), the volume of the influencing stream, the biosolids-to-water ratio and the sludge retention time all affect the micropollutant distribution in the different phases. It is also possible that there is a classification: biodegradable micropollutants are transformed into finished products (CO_2, H_2O), micropollutants that are difficult to biodegrade depend on operating conditions, while refractory micropollutants are eliminated by adhesion to biomass. Observations at one treatment plant are not systematically reproduced at another, even if the plant is supplied with the same micropollutant load. The appearance of micropollutants in various streams of municipal wastewater treatment plants, as well as their fate and elimination processes, are here discussed. The fate of micropollutants in wastewater treatment plants depends largely on adsorption onto suspended particles, primary and secondary sludge and dissolved organic carbon, with removal occurring through coagulation–flocculation and biodegradation. The log Kow (> 2.5) and pKa are the dominant micropollutant properties, and the concentration, organic fraction and surface loading of suspended particles dictate the degree of micropollutant adsorption. Most conventional wastewater treatment plants do not effectively remove complex micropollutants through biodegradation or biotransformation ($k_{bio} \leq 0.0042$ L/gss/h). Elimination varies considerably for different compounds, as well as for the same substance, due to operating conditions such as aerobic, anaerobic and anoxic conditions;

sludge retention time; pH; redox potential; and temperature. Membrane bioreactors perform better for moderately biodegradable compounds due to the diverse nature of microorganisms and greater adaptability resulting from a longer sludge retention time.

Filamentous bacteria contribute to the adhesion of small flocs to form a more robust floc macrostructure, but the flocculation process is essentially due to electrostatic interactions between microorganisms, exopolymers and metal ions. Bonds form between the negatively charged functional groups of exopolymers and divalent cations such as Ca^{2+} or Mg^{2+}. This bridging promotes aggregation and stabilizes the matrix formed by biopolymers and microorganisms.

Extracellular polymers (ECPs) play a key role in the adsorption of metallic micropollutants (Guibaud et al. 2009). Furthermore, humic acids (insoluble at $pH < 2$ and soluble at higher pH levels) and fulvic acids (soluble at all pH levels) are substances that play an important role in micropollutant interaction. This humic material is generally considered to be the constituent with the greatest influence on the transport and behavior of organic micropollutants present in organic matrices such as sludge (Pan et al. 2008).

Due to their physicochemical properties (molecular weight, solubility, water–octanol partition coefficient Kow, partition coefficient with organic carbon Koc, etc.), many micropollutants are likely to be degraded from the water stream during wastewater treatment, and byproducts end up in the effluent. Some of them are adsorbed onto bacterial flocs and then end up in the extracted sludge.

At low concentrations, the micropollutant biodegradation/biotransformation kinetics follow first-order reaction kinetics:

$$\text{Rate} = \text{Kbio X Co}$$

where Kbio is the biodegradation rate constant, X is the suspended solid concentration in the reactor and Co is the micropollutant concentration in the wastewater.

As a general rule, complex aromatic structures containing more than one benzene ring and/or chlorine and nitro groups are not efficiently biodegraded (Pomiès 2013). The aerobic biodegradation constants of 20 aromatic species using activated sludge were reported, and the kinetic constants were correlated with the molecular structure (Andreozzi et al. 2006). The normalized first-order rate Kbio constants ($L.g^{-1}TSS.h^{-1}$) using X ($g.L^{-1}$) were 0.003, 0.02 and 3.80, respectively, for 3,5-dinitrobenzoic acid, 2,6-dichlorophenol and benzoic acid.

Pomiès (2013) summarized a list of aerobic and anaerobic rate constants for 20 pharmaceuticals, including antibiotics and compounds such as bisphenol A and nonylphenol. The aerobic rate constants Kbio range from 0.0025 to 7.08 $L.g^{-1}SS.h^{-1}$. Carbamazepine had the lowest value, while galaxolide proved to be the most biodegradable compound. Log Kow was 2.45 for carbamazepine and 5.9 for galaxolide. Compounds with Kbio values < 0.0042 $L.g^{-1}SS.h^{-1}$ are not well removed (< 20%), whereas compounds with Kbio > 0.4 $L.g^{-1}SS.h^{-1}$ are transformed at a rate of over 90%. Consequently, with existing biological treatment schemes in municipal wastewater, 90% of micropollutants are not removed or biotransformed. Much plant data do not distinguish between adsorption and biotransformation due to chemical analysis challenges. In most cases, overall elimination is estimated based on the concentrations of the influencer and the effluent, and information on intermediate steps is not systematic. Many questions remain unanswered, such as the fate of intermediate metabolites, pharmaceutical processing products and the complex chemistry involved in these compounds.

In fact, removal efficiency depends heavily on the physicochemical properties of a substance. The removal of hydrophobic compounds occurs first during the removal of suspended solids, in the primary treatment stage. Micropollutants can also be removed during secondary treatment, such as in activated sludge, by sorption onto biological sludge and biodegradation. Furthermore, removal efficiencies vary depending on the operating conditions of the wastewater treatment plant, such as sludge retention time, hydraulic retention time and temperature – although the influence of these parameters is not always clearly understood. For example, higher removal efficiencies were observed for most of the micropollutants studied when nitrification occurs due to a high sludge retention time (Choubert et al. 2011). Previous studies have shown that a higher sludge retention time increases the elimination efficiency of pharmaceutical compounds (Clara et al. 2004; Carucci et al. 2006). Indeed, a longer sludge retention time could promote the diversity of bacterial communities, as well as the presence of slower-growing species, thus increasing the biomass biodegradation potential (Kreuzinger et al. 2004; Suarez et al. 2010). Furthermore, a high sludge retention time combined with a reduced food/microorganism ratio appears to promote antibiotic biodegradation (Göbel et al. 2007). However, some authors have not observed any improvement in the elimination efficiency of pharmaceutical compounds for a sludge retention time between 10 and 80 days (Joss et al. 2005; Vieno et al. 2007). A longer hydraulic retention time implies a longer contact time between the wastewater and the sludge, which appears to promote the biodegradation reaction. Evaluating the degradability of micropollutants, in terms of half-lives, is an important task for discussing integrated solutions for mitigating pollutant entry into the water cycle. By linking the calculation of micropollutant removal rates to half-lives, while assuming that compound degradation follows pseudo-first-order kinetics, it is suggested that conventional wastewater treatment processes applied to wastewater treatment plants are incapable of completely removing all pharmaceuticals. The long half-lives observed by many pharmaceuticals lead to the use of longer hydraulic retention times to improve their degradation.

Understanding the influence of operating conditions on the removal of micropollutants is key to better understanding and improving removal mechanisms. However, as shown above, the experimental results presented in the literature do not allow for definite conclusions.

In fixed-film processes, biofilms offer several advantages over free-culture processes for removing micropollutants: a high biomass retention time, high microbial diversity, the presence of microniches and redox potential conditions are biofilm characteristics that promote micropollutant removal.

The biofilm harbors a significant diversity of microorganisms that can be very beneficial in eliminating micropollutants.

Drug residues, pesticides and hormones are effectively eliminated and are quantified at levels often below 0.1 μg.L^{-1} or even at values below the limit of detection (LOD). However, polycyclic aromatic hydrocarbons (PAHs), detergents, plasticizers and certain chemicals are quantified in the sludge and at residual concentrations in the effluent (Ellis 2006; Kümmerer 2009; Besnault and Martin 2011).

The concentrations of micropollutants quantified in the sludge are shown in Figure 1.11.

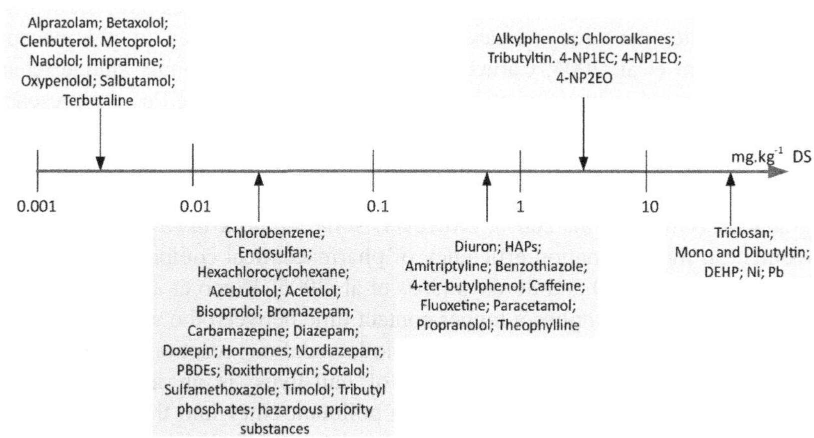

Figure 1.11. *Average concentrations of various micropollutants quantified in the extracted sludge (Besnault et al. 2009; Besnault and Martin 2011)*

Pesticides and insecticides are not frequently found in sludge, with the exception of diuron, alachlor and aldrin, which are found at average concentrations between 0.1 and 1 mg.kg $^{-1}$SS in French wastewater treatment plant sludge (Coquery et al. 2011).

With regard to medicines (or pharmaceutical compounds) in particular, the effectiveness of the wastewater treatment plants is variable and not absolute (Heberer 2002; Petrovic et al. 2003; Yu et al. 2006). Upon exiting the treatment plant, a fraction of the initial wastewater contamination remains, consisting of pharmaceutical compounds that can contaminate natural waters. Among these compounds, the removal of antibiotics is of particular interest because they can lead to antibiotic resistance (Stalder 2012).

Overall, drug residues are found in natural waters at concentrations of approximately 0.01–1 μg/L (Heberer 2002; Lindberg et al. 2005; Yu et al. 2006; Mullot et al. 2010).

Due to their physicochemical properties, drugs for human and veterinary use such as ciprofloxacin, erythromycin, ibuprofen, tetracyclines and fluoroquinolones are preferentially adsorbed onto bacterial flocs (Beausse 2004). Levels exceeding 0.1 mg.kg^{-1}SS traces of fluoxetine, paracetamol, propranolol or amitriptyline were found in the sludge (Soulier et al. 2011).

The 2009 EPA study measured more than 70 pharmaceutical products in the sludge and found low levels for most substances, such as trimethoprim (0.012–0.2 mg.kg^{-1}SS) or clarithromycin (0.0087–0.62 mg.kg^{-1}SS). Other substances were found at higher levels, such as carbamazepine (0.009–6.03 mg.kg^{-1}SS) or cimetidine (0.004–8.33 mg.kg^{-1}SS). For antibiotics, the most common groups found in the environment are macrolides, fluoroquinolones and sulfonamides, while tetracyclines and penicillins are generally found at lower concentrations (Golet et al. 2003; Beausse 2004; Lindberg et al. 2006).

In Switzerland, fluoroquinolones (in particular ciprofloxacin and norfloxacin) are molecules that have been detected in both wastewater and sludge (Golet et al. 2002). They were found there at levels in the order of ng.L^{-1} in water and mg.kg^{-1} in sewage sludge. The presence of fluoroquinolones (norfloxacin and ciprofloxacin) has been measured in wastewater and sludge from treatment plants in Sweden (Lindberg 2006). Concentrations of 66 ng/L of norfloxacin and 74 ng/L of ciprofloxacin were measured in filtered wastewater. In sludge, ciprofloxacin concentrations are higher: 2.4 mg.kg^{-1}SS (activated sludge) and 6.3 mg.kg^{-1}SS (digested sludge).

Table 1.6 lists the removal rates of pharmaceutical compounds in wastewater treatment plants.

Hormones	Process (min–max removal percentage)				
	Activated sludge	Trickling filters	Biofilters	MBR	MBBR
Estrone	55–79	58–63	84–95	94–95	–
17-β-estradiol	69–99	64–70	85–96	90–96	92–96
17-α-estradiol	69–98	72–76	81–94	91–94	90–94
Ethinylestradiol	74–94	60–83	34–58	64–78	–

Table 1.6. *Percentage of hormone elimination by biological processes*

Substances	BRM process		
	Inlet ($ng.L^{-1}$)	Outlet ($ng.L^{-1}$)	Effectiveness (%)
Paracetamol	28,370	356	98
Diclofenac	1,494	808	46
Fluoxetine	1,527	1,067	30
Carbamazepine	422	126	70
Bezafibrate	475	146	70
Gemfibrozil	888	161	81
Metoprolol	224	< LOQ	98
Propranolol	328	210	57
Sulfamerazine	52	< LOQ	LOD
Sulfamethoxazole	480	141	70
Trimethoprim	246	119	51
Clarithromycin	540	49	90
Erythromycin	66	13	> 80

Table 1.7. *BRM effectiveness (Biosep™ Veolia) in micropollutant removal (LOQ: limit of quantification, LOD: limit of detection)*

Thus, for many micropollutants, it appears extremely difficult to predict or generalize the removal efficiency values of activated sludge, which can vary greatly from one installation to another (Miège et al. 2009).

1.4.2. *Activated sludge*

Micropollutant		Plant inlet	Plant outlet	Reliability	Amperes
Organic micropollutants	Estradiol-alpha ng.L^{-1}	12	< 0.4	LOQ	90–100%
	Estradiol-beta ng.L^{-1}	40	0.7	–	80–100%
	Estrone ng.L^{-1}	96	3.3	–	80–100%
	Ethinylestradiol ng.L^{-1}	< 0.8	< 0.8	LOQ	–
Benzene compounds	Toluene µg.L^{-1}	2.2	< 0.5	OK	–
Polycyclic aromatic hydrocarbons	Benzo (b) fluoranthene µg.L^{-1}	0.033	< 0.005	LOQ	70–100%
	Benzo (k) fluoranthene	0.02	< 0.005	LOQ	80–90%
	Benzo (g, h, j) perylene µg.L^{-1}	0.017	< 0.005	LOQ	–
	Indeno (1,2,3-c,d) pyrene µg.L^{-1}	0.034	< 0.005	LOQ	80–90%
	Fluoranthene µg.L^{-1}	0.1	< 0.01	LOQ	70–100%
	Benzo (a) pyrene (3,4) µg.L^{-1}	0.03	< 0.01	LOQ	–
Urea-based pesticides	Diuron µg.L^{-1}	0.04	0.3	LOD	0–50%
Non-ionic detergents	4-Nonylphenol-diethoxylate (NP2OE) µg.L^{-1}	< 0.1	< 0.1	LOQ	–
	4-Nonylphenol-ethoxylate (NP1OE) µg.L^{-1}	0.4	< 0.1	LOQ	–
	Octylphenol-ethoxylates (OP1OE) µg.L^{-1}	< 0.1	< 0.1	LOQ	–
	Octylphenol-diethoxylates (OP2OE) µg.L^{-1}	< 0.1	< 0.1	LOQ	–
	Octylphenols µg.L^{-1}		0.043	LOD	70–100%
	Nonylphenols µg.L^{-1}	0.2	1.757	LOD	60–100%
Analgesics/anti-inflammatories	Paracetamol ng.L^{-1}	> 2,500	1,006	LOQ	90–100%
	Diclofenac ng.L^{-1}	2,509	2,509	LOD	0–40%
	Phenazone ng.L^{-1}	< 25	< 25	LOQ	–
Antidepressant	Fluoxetine ng.L^{-1}	21	32	LOD	15–85%

Micropollutant		Plant inlet	Plant outlet	Reliability	Amperes
Anti-epileptic	Carbamazepine ng.L^{-1}	395	390	OK	0%
	Primidone ng.L^{-1}	461	462	LOD	–
Liporegulators	Bezafibrate ng.L^{-1}	351	157	OK	–
	Gemfibrozil ng.L^{-1}	422	153	OK	70–100%
Beta-blockers	Metoprolol ng.L^{-1}	492	380	OK	30–100%
	Propranolol ng.L^{-1}	1,092	1,060	OK	0–40%
Antibiotics	Sulfamerazine ng.L^{-1}	< 25	< 25	LOQ	–
	Sulfamethoxazole ng.L^{-1}	3,106	1,004	OK	30–70%
	Sulfachloropyridazine ng.L^{-1}	< 25	< 25	LOQ	–
	Trimethoprim ng.L^{-1}	282	215	OK	–
Macrolide antibiotics	Erythromycin + erythromycin-H_2O ng.L^{-1}	143.6	118	OK	–
	Clarithromycin ng.L^{-1}	54.2	74.1	LOD	–
	Roxithromycin ng.L^{-1}	162.1	274.9	LOD	20–60%

Table 1.8. *Biological efficiency using activated sludge on the reduction of micropollutants (LOQ: limit of quantification; LOD: limit of detection)*

1.5. Biological removal of micropollutants using hybrid systems

Fixed-film processes result in longer sludge ages than free-film processes, as well as potentially higher biodiversity. Hybrid systems with a moving support (MBBRs) appear to be a promising alternative to conventional activated sludge for the removal of poorly biodegradable micropollutants such as diclofenac and contrast agents. Biofilm thickness can potentially impact biomass structure and activity. The diffusion of substrates, particularly oxygen, into the biofilm is the primary load-limiting process in a biomass bioreactor (MBBR), creating substrate gradients within the biofilm. Its thickness induces steeper concentration gradients and a stratification of metabolic processes, resulting in the formation of a more heterogeneous and diverse biofilm (Torresi et al. 2016; Edefell et al. 2021).

However, it is not yet known exactly how biofilm thickness influences biodiversity and functionality (e.g. micropollutant removal) in biofilm systems, partly due to the lack of technologies to control biofilm thickness.

In order to obtain biofilms of varying thicknesses, researchers at the company Veolia-AnoxKaldnes™ used Z-shaped polyethylene supports with a saddle-shaped grid surface, allowing biofilm to grow on the outside the support rather than inside it (see Figure 1.12). Because the supports continuously rub against each other during MBBR operation, the grid thickness corresponds to the maximum biofilm thickness.

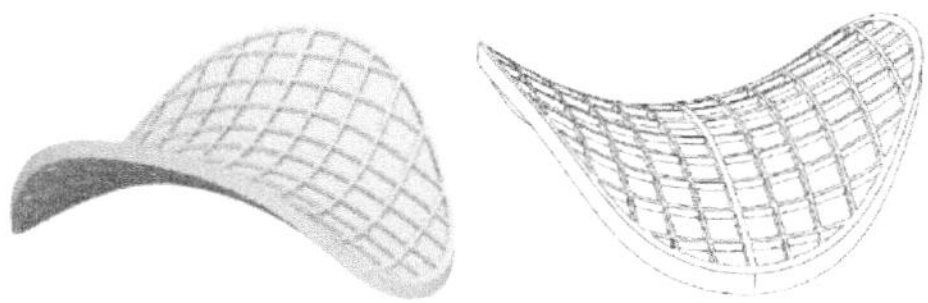

Figure 1.12. *Z supports, AnoxKaldnes™*

The principle of biological treatment using the Exeno™-M process is shown in Figure 1.13.

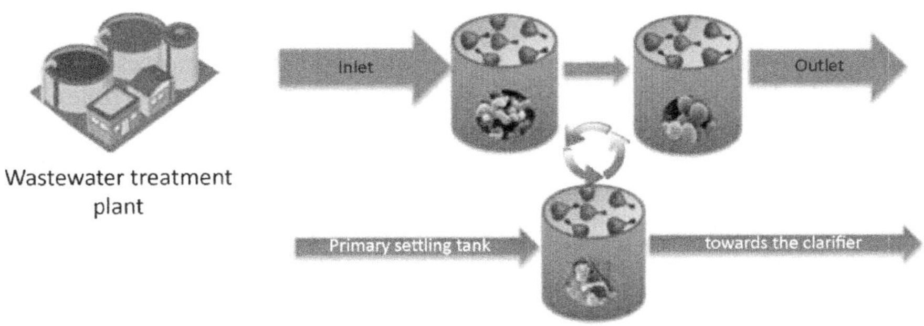

Figure 1.13. *Principle of the Exeno™-M process (AnoxKaldnes™)*

The biotransformation mechanism is governed by a pseudo-first-order equation. The rate constant kbio ($L.g^{-1}$ SS. d^{-1}) is obtained by:

Cf = Co e-Xt

where Cf is the concentration of the micropollutant (ng L^{-1}) measured in the reactor and X is the biomass concentration on the Z support ($g.L^{-1}$). Sorption on the biofilm is represented by the coefficient Kd ($L.g^{-1}$). At equilibrium, we write:

$$\frac{dC}{dt} = -\frac{kbio\ CX}{(1 + KdX)}$$

Retransformation rates kdec ($L.g^{-1}$ SS.d) are estimated using the equation:

$$\frac{dC}{dt} = -\frac{kbio\ CX}{(1 + KdX)} + kdec\ CX$$

$$kbio = k/X$$

$$ksurf = k/A$$

where C is the dissolved concentration of the compound ($ng.L^{-1}$), k is the biodegradation rate constant (d^{-1}), t is time (d), X is the concentration of biomass on the support ($g.L^{-1}$), kbio is the biodegradation rate constant relative to biomass ($L.g^{-1}$ $SS.d^{-1}$) and ksurf is the degradation rate constant relative to the total surface area of the supporting materials ($L.m^{-2}$. d^{-1}).

Figure 1.14 shows the biodegradation rate constants per unit area for biofilms in MBBRs with high secondary and tertiary loading. The differences are more pronounced for sulfonamides (sulfamethoxazole and sulfamethiazole) and nonsteroidal anti-inflammatory drugs (ibuprofen and phenazone). As in activated sludge, an increase in sulfonamide biodegradation is observed when the MBBR is operating at a high loading (Majewsky et al. 2011; Achermann et al. 2018b). This is linked to strong bacterial growth and folic acid synthesis (Achermann et al. 2018a).

However, substantial differences are evident between secondary and tertiary treatments under high-load conditions. These results cannot be explained solely by the difference in biomass concentrations. The biodegradation capacity of several micropollutants, such as beta-blockers (atenolol and metoprolol), antibiotics (clarithromycin and erythromycin) and iodinated contrast agents (iohexol, iomeprol and iopromide), is more concentrated when comparing rate constants per unit biomass, kbio. The tertiary stage results are more significant than those for the high-loading stage. Therefore, decreased substrate availability appears to have a positive effect on kbio for several micropollutants, but at the cost of a decrease in biomass concentration.

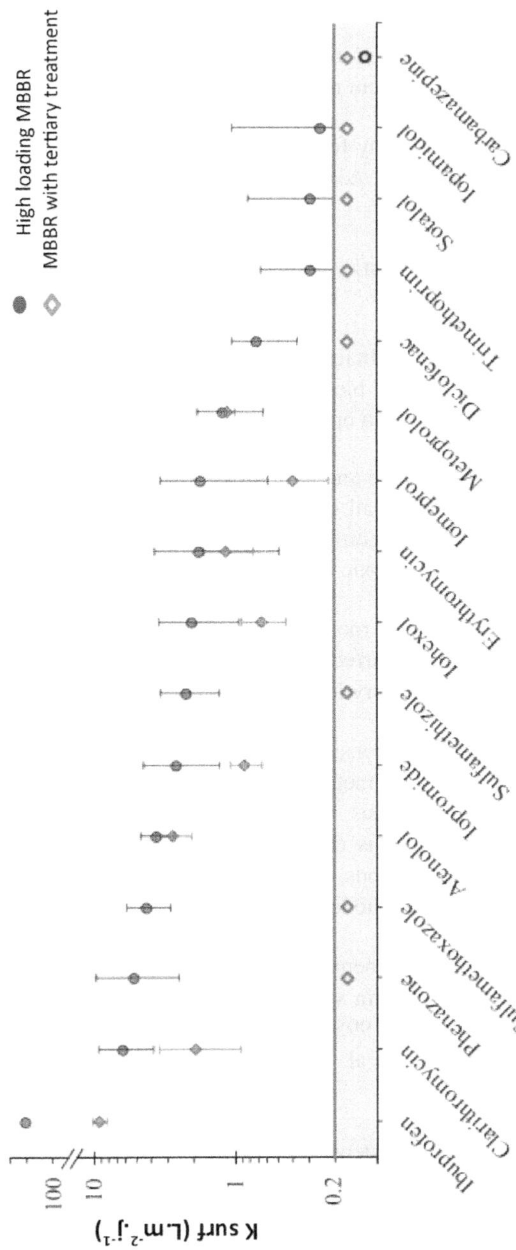

Figure 1.14. *Biodegradation rate constant ksurf, relative to the total surface area of the supports (Edefell et al. 2021). For a color version of this figure, see www.iste.co.uk/gaid/wastewaterengineering4.zip*

1.5.1. *Impact of effluent recirculation*

A systematic increase is observed in biomass concentration and ksurf values during the operation of MBBRs with effluent recirculation.

ksurf values increase significantly for iodinated contrast agents (iopromide, iohexol, iopamidol and iomeprol) when an additional substrate is added to the biofilm in a recirculation configuration.

This solution exhibits higher ksurf values for atenolol and metoprolol compared to a conventional MBBR regime.

Figure 1.15 shows the results obtained in MBBRs operating under aerobic, anoxic and anaerobic conditions. Higher biodegradation rates are obtained under aerobic conditions compared to the other two operating modes.

Trimethoprim, sulfamethoxazole and antibiotics (clarithromycin, roxithromycin and erythromycin) are degraded under all conditions (see Figure 1.15). Sulfamethoxazole and trimethoprim are the only compounds with higher degradation rates under anaerobic conditions than under anoxic conditions (Edefell et al. 2021).

Iodinated contrast agents are more sensitive to redox than trimethoprim and antibiotics. Their degradation occurred at higher rates under aerobic conditions than under anoxic conditions and was very low under anaerobic conditions.

Beta-blockers exhibit variable sensitivity to redox conditions. This sensitivity is as pronounced for atenolol as it is for metoprolol and sotalol. Atenolol degrades under all conditions, likely due to hydrolysis of the primary amide (Helbling et al. 2010; Achermann et al. 2018b). Sotalol is degradable under aerobic conditions, but stable under anoxic and anaerobic conditions. The same is true for other micropollutants such as ibuprofen, diclofenac, valsartan and losartan.

The molecular oxygen requirements for the degradation of many micropollutants are particularly concerning in biofilm systems, where limiting diffusion within biofilm structures is crucial (Bishop et al. 1995), leading to a stratification of microbial layers (Almstrand et al. 2014; Suarez et al. 2019) and to changes in metabolic pathways (Alvarino et al. 2018b).

In a stratified biofilm, only the aerated outer layer would be likely to contribute to the aerobic biodegradation of micropollutants. This may justify the choice of considering the parameter ksurf rather than kbio.

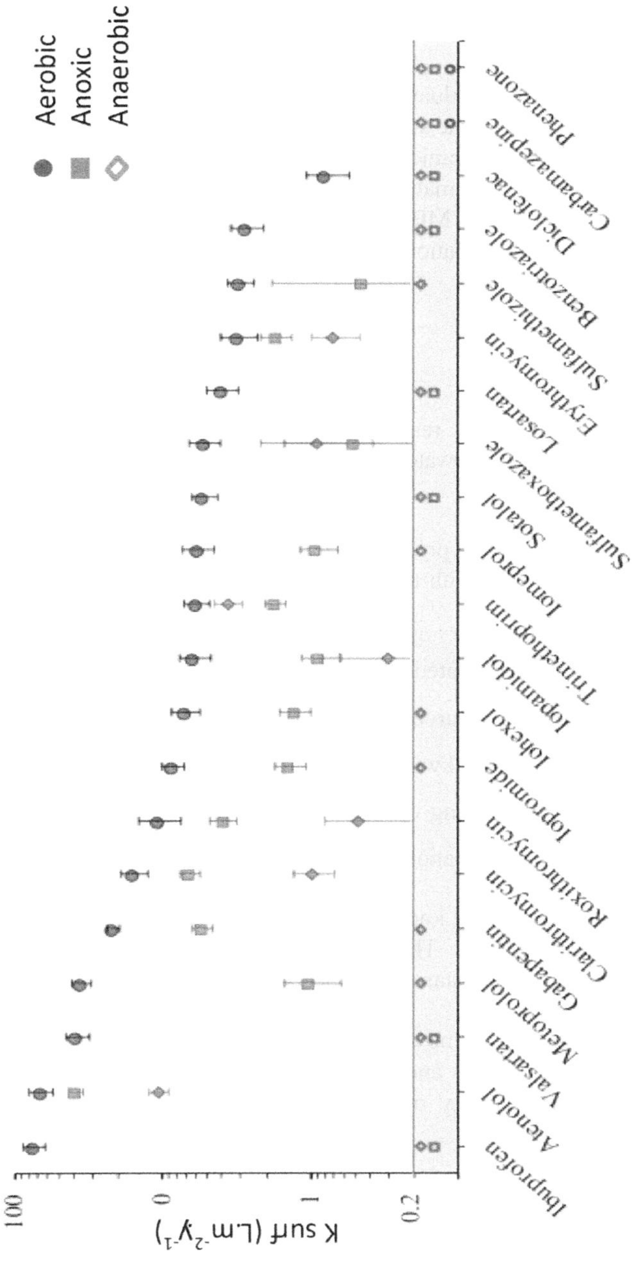

Figure 1.15. *Biodegradation rate constant ksurf under different operating conditions (aerobic: presence of oxygen; anoxia: presence of nitrates; anaerobic: without oxygen or nitrates) (Edefell et al. 2021). For a color version of this figure, see www.iste.co.uk/gaid/wastewaterengineering4.zip*

A similar line of reasoning applies to nitrification in biofilms, as nitrification rates are negatively affected by oxygen depletion within the biofilm caused by high organic loads

It thus appears that the evaluation of the biodegradation rate constants of micropollutants as a function of the support surface area in a high-load MBBR and a tertiary MBBR shows that the organic load is an important parameter due to the high biomass that accumulates on the materials. This is confirmed if a portion of the raw wastewater supplies the tertiary MBBR and influences biomass growth. Aerobic conditions promote the biodegradation of many micropollutants, such as diclofenac, metoprolol and valsartan.

1.6. Conclusion

The current challenge is to remove micropollutants and microplastics from wastewater, especially if the wastewater is intended for reuse as drinking water (urban and industrial).

Therefore, additional treatment processes are necessary (Gaid 2025a, 2025b). Some factories are already operational with these additional treatments (known as quaternary treatments).

Micropollutants can be eliminated using various methods:

– adsorption onto activated charcoal (powder or granules);

– ozonation alone or combined with activated charcoal;

– advanced oxidation combining various chemical and physical oxidants;

– high-pressure membranes (nanofiltration and reverse osmosis).

These treatments are now well known because most have already been tested and used for drinking water treatment. They have proven their effectiveness in the field against organic matter (humic substances), as described in Gaid (2025a, 2025b).

As mechanisms for the removal of organic pollutants, the efficiency of transformation by biological units and the role of microbial communities are poorly understood, and the predictability of trace organic contaminant removal is very limited.

1.7. References

Achermann, S., Bianco, V., Mansfeldt, C.B., Vogler, B., Kolvenbach, B.A., Corvini, P.F.X., Fenner, K. (2018a). Biotransformation of sulfonamide antibiotics in activated sludge: The formation of pterin-conjugates leads to sustained risk. *Environ. Sci. Technol.*, 52(11), 6265–6274.

Achermann, S., Falås, P., Joss, A., Mansfeldt, C.B., Men, Y., Vogler, B., Fenner, K. (2018b). Trends in micropollutant biotransformation along a solids retention time gradient. *Environ. Sci. Technol.*, 52(20), 11601–11611.

ADEME (1995). *Les micropolluants métalliques dans les boues résiduaires des stations d'épuration urbaines.* ADEME Editions, Paris.

ADEME (1999). Épandages des boues d'épuration sur prairies et cultures fourragères : aspects. Report, Microbiologiques.

Almstrand, R., Persson, F., Daims, H., Ekenberg, M., Christensson, M., Wilen, B.-M., Sorensson, F., Hermansson, M. (2014). Three-dimensional stratification of bacterial biofilm populations in a moving bed biofilm reactor for nitritation–anammox. *Int. J. Mol. Sci.*, 15(2), 2191–2206.

Alvarino, T., Suarez, S., Lema, J., Omil, F. (2018a). Understanding the sorption and biotransformation of organic micropollutants in innovative biological wastewater treatment technologies. *Sci. Total Environ.*, 615, 297–306.

Alvarino, T., Lema, J., Omil, F., Suarez, S. (2018b). Trends in organic micropollutants removal in secondary treatment of sewage. *Rev. Environ. Sci. Biotechnol.*, 17, 447–469.

Andreozzi, R., Raffaele, M., Nicklas, P. (2003). Pharmaceuticals in STP effluents and their solar photodegradation in aquatic environment. *Chemosphere*, 50, 1319–1330.

Andreozzi, R., Cesaro, R., Marotta, R., Pirozzi, F. (2006). Evaluation of biodegradation kinetic constants for aromatic compounds by means of aerobic batch experiments. *Chemosphere*, 2(9), 1431–1436.

Barret, M., Carrere, H., Latrille, E., Wisniewski, C., Patureau, D. (2010a). Micropollutant and sludge characterization for modeling sorption equilibria. *Env. Sci. Technol.*, 44, 1100–1106.

Barret, M., Patureau, D., Latrille, E., Carrère, H. (2010b). A three-compartment model for micro-pollutants sorption in sludge: Methodological approach and insights. *Water Res.*, 44, 616–624.

Barret, M., Delgadillo-Mirquez, L., Trably, E., Delgenès, N., Braun, F., Cea-Barcia, G., Steyer, J.P., Patureau, D. (2012). Anaerobic removal of trace organic contaminants in sewage sludge: 15 years of experience. *Pedosphere*, 22, 508–517.

Beausse, J. (2004). Selected drugs in solid matrices: A review of environmental determination occurrence and properties of principal substances. *Trends Anal. Chem.*, 23, 753–761.

Besnault, S. and Martin, S. (2011). Etat de l'art sur les procedes de traitement des boues pour l'elimination de micropolluants. ARMISTIQ – Action D.

Besnault, S., Martin-Ruel, S., Baig, S., Heiniger, B., Esperanza, M., Budzinski, H., Miege, C., Menach, K. , Dherret, L., Roussel-Galle, A. et al. (2015). Évaluation technique, économique et environnementale de procédés de traitement complémentaire avancés pour l'élimination des micropolluants. *Tech. Sci. Méthodes*, 3, 67–83.

Bishop, P.L., Zhang, T.C., Fu, Y.C. (1995). Effects of biofilm structure, microbial distributions and mass-transport on biodegradation processes. *Water Sci. Technol.*, 31(1), 143–152.

Boxall, A.B.A., Sinclair, C.J., Fenner, K., Kolpin, D., Maund, S.J. (2004). Peer reviewed: When synthetic chemicals degrade in the environment. *Environ. Sci. Technol.*, 38, 368A–375A.

Burzio, C. (2022). Removal of micropollutants from wastewater in aerobic granular sludge and activated sludge systems. PhD Thesis, Chalmers University of Technology, Gothenburg.

Carucci, A., Cappai, G., Piredda, M. (2006). Biodegradability and toxicity of pharmaceuticals in biological wastewater treatment plants. *J. Environ. Sci. Health A Tox. Hazard Subst. Environ. Eng.*, 41, 1831–1842.

Chipasa, K.B. (2003). Accumulation and fate of selected heavy metals in a biological wastewater treatment system. *Waste Manag.*, 23, 135–143.

Choubert, J.-M., Martin-Ruel, S., Budzinski, H., Miège, C., Esperanza, M., Soulier, C., Ugarrigue, C., Coquery, M. (2011). Removal of micropollutants by domestic conventional wastewater treatment plants and advanced tertiary process: Specific method and results of the Ampères project. *Tech. Sci. Méthodes*, 1/2, 44–62.

Chu, C.P. and Lee, D.J. (2004). Multiscale structures of biological flocs. *Chem. Eng. Sci.*, 59, 1875–1883.

Clara, M., Strenn, B., Saracevic, E., Kreuzinger, N. (2004). Adsorption of bisphenol-A, 17β-Estradiol and 17α-ethinylestradiol to sewage sludge. *Chemosphere*, 56, 843–851.

Clara, M., Gans, O., Windhofer, G., Krenn, U., Hartl, W., Braun, K., Scharf, S., Scheffknecht, C. (2011). Occurrence of polycyclic musks in wastewater and receiving water bodies and fate during wastewater treatment. *Chemosphere*, 82(8), 1116–1123.

Coquery, M., Pomiès, M., Martin-Ruel, S., Budzinski, H., Miège, C., Esperanza, M., Soulier, C., Choubert, J.-M. (2011). Concentrations and fluxes of micropolluants in wastewaters and sludge: Methodology and main results of the Amperes project. *Tech. Sci. Méthodes*, 1/2, 25–43.

Deblonde, T., Cossu-Leguille, C., Hartemann, P. (2011). Emerging pollutants in wastewater: A review of the literature. *Int. J. Hyg. Environ. Health*, 214, 442–448.

Edefell, E., Falås, P., Torresi, E., Hagman, M., Cimbritz, M., Bester, K., Christensson, M. (2021). Promoting the degradation of organic micropollutants in tertiary moving bed biofilm reactors by controlling growth and redox conditions. *J. Hazard. Mater.*, 414, 1–8.

Eikelboom, D.H. (2000). *Process Control of Activated Sludge Plants by Microscopic Investigation*. International Water Association (IWA) Publishing, London.

Ellis, J.B. (2006). Pharmaceutical and personal care products (PPCPs) in urban receiving waters. *Environ. Pollut.*, 144, 184–189.

Escolà Casas, M., Nielsen, T.K., Kot, W., Hansen, L.H., Johansen, A., Bester, K. (2017). Degradation of mecoprop in polluted landfill leachate and waste water in a moving bed biofilm reactor. *Water Res.*, 121, 213–220.

Falås, P., Andersen, H.R., Ledin, A., La Cour Jansen, J. (2012a). Occurrence and reduction of pharmaceuticals in the water phase at Sweden wastewater treatment plants. *Water Sci. Technol.*, 66, 783–791.

Falås, P., Baillon-Dhumez, A., Andersen, H.R., Ledin, A., La Cour Jansen, J. (2012b). Suspended biofilm carrier and activated sludge removal of acidic pharmaceuticals. *Water Res.*, 46, 1167–1175.

Falås, P., Longrée, P., La Cour Jansen, J., Siegrist, H., Hollender, J., Joss, A. (2013). Micro-pollutant removal by attached and suspended growth in a hybrid biofilm-activated sludge process. *Water Res.*, 47, 4498–4506.

Falås, P., Wick, A., Castronovo, S., Habermacher, J., Ternes, T.A., Joss, A. (2016). Tracing the limits of organic micropollutant removal in biological wastewater treatment. *Water Res.*, 95, 240–249.

Feng, S., Tan, C.H., Constancias, F., Kohli, G.S., Cohen, Y., Rice, S.A. (2017). Predation by Bdellovibrio bacteriovirus significantly reduces viability and alters the microbial community composition of activated sludge flocs and granules. *FEMS Microbiol. Ecol.*, 93, 128–139.

Fischer, K. and Majewsky, M. (2014). Co-metabolic degradation of organic wastewater micropol-lutants by activated sludge and sludge inherent microorganisms. *Appl. Microbiol. Biotechnol.*, 98(15), 6583–6597.

Franco, A. and Trapp, S. (2008). Estimation of the soil–water partition coefficient normalized organic carbon for ionizable organic chemicals. *Environ. Toxicol. Chem.*, 27, 1995–2004.

Gaid, K. (2025a). *Wastewater Reuse 1: Characteristics, Uses, Applications, Filtration and Disinfection of Water*. ISTE Ltd, London, and Wiley, Hoboken, NJ.

Gaid, K. (2025b). *Wastewater Reuse 2: Micropollutants, Membranes and Treatment Procedures*. ISTE Ltd, London, and Wiley, Hoboken, NJ.

García-Galán, J.J., Anfruns, A., Gonzalez-Olmos, R., Rodríguez-Mozaz, S., Comas, J. (2016). UV/H_2O_2 degradation of the antidepressants venlafaxine and O-desmethylvenlafaxine: Elucidation of their transformation pathway and environmental fate. *J. Hazard. Mater.*, 311, 70–80.

Göbel, A., Thomsen, A., McArdell, C.S., Joss, A., Giger, W. (2005). Occurrence and sorption behaviour of sulfonamides, macrolides, and trimethoprim in activated sludge treatment. *Env. Sci. Technol.*, 39, 3981–3989.

Göbel, A., McArdell, C.S., Joss, A., Siegrist, H., Giger, W. (2007). Fate of sulfonamides, macrolides, and trimethoprim in different wastewater treatment technologies. *Sci. Tot. Environ.*, 372(2–3), 361–371.

Golet, E.M., Alder, A.C., Hartmann, A., Ternes, T.A., Giger, W. (2001). Trace determination of fluoroquinolone antibacterial agents in urban wastewater by solid-phase extraction and liquid chromatography with fluorescence detection. *Anal. Chem.*, 73, 3632–3638.

Golet, E.M., Alder, A.C., Giger, W. (2002). Environmental exposure and risk assessment of fluoroquinolone antibacterial agents in wastewater and river water of the Glatt Valley water shed, Switzerland. *Environ. Sci. Technol.*, 36, 3645–3651.

Golet, E.M., Xifra, I., Siegrist, H., Alder, A.C., Giger, W. (2003). Environmental exposure assessment of fluoroquinolone antibacterial agents from sewage to soil. *Environ. Sci. Technol.*, 37, 3243–3249.

Gros, M., Petrović, M., Ginebreda, A., Barceló, D. (2010). Removal of pharmaceuticals during wastewater treatment and environmental risk assessment using hazard indexes. *Environ. Int.*, 36, 15–26.

Guibaud, G., van Hullebusch, E., Bordas, F., d'Abzac, P., Joussein, E. (2009). Sorption of Cd (II) and Pb (II) by exopolymeric substances (EPS) extracted from activated sludge and pure bacterial strains: Modeling of the metal/ligand ratio effect and role of the mineral fraction. *Bioresour. Technol.*, 100, 2959–2968.

Halling-Sørensen, B., Nors Nielsen, S., Lanzky, P.F., Ingerslev, F., Holten Lützhøft, H.C., Jørgensen, S.E. (1998). Occurrence, fate and effects of pharmaceutical substances in the environment – A review. *Chemosphere*, 36, 357–393.

Heberer, T. (2002). Occurrence, fate, and removal of pharmaceutical residues in the aquatic environment: A review of recent research data. *Toxicol. Lett.*, 131, 5–17.

Helbling, D.E., Hollender, J., Kohler, H.-P.E., Fenner, K. (2010). Structure-based interpretation of biotransformation pathways of amide-containing compounds in sludge-seeded bioreactors. *Environ. Sci. Technol.*, 44(17), 6628–6635.

Hermansson, M. (1999). The DLVO theory in microbial adhesion. *Colloids Surf. Biointerfaces*, 14, 105–119.

Hirsch, R., Ternes, T.A., Haberer, K., Mehlich, A., Ballwanz, F., Kratz, K.-L. (1998). Determination of antibiotics in different water compartments via liquid chromatography – Electrospray tandem mass spectrometry. *J. Chromatogr.*, 815, 213–223.

Howard, P.H. and Muir, D.C.G. (2010). Identifying new persistent and bioaccumulative organics among chemicals in commerce. *Environ. Sci. Technol.*, 44, 2277–2285.

Hyland, K.C., Dickenson, E.R.V., Drewes, J.E., Higgins, C.P. (2012). Sorption of ionized and neutral emerging trace organic compounds onto activated sludge from different wastewater treatment configurations. *Water Res.*, 46, 1958–1968.

Joss, A., Keller, E., Alder, A.C., Göbel, A., McArdell, C.S., Ternes, T., Siegrist, H. (2005). Removal of pharmaceuticals and fragrances in biological wastewater treatment. *Water Res.*, 39, 3139–3152.

Joss, A., Zabczynski, S., Göbel, A., Hoffmann, B., Löffler, D., McArdell, C.S., Ternes, T.A., Thomsen, A., Siegrist, H. (2006). Biological degradation of pharmaceuticals in municipal wastewater treatment: Proposing a classification scheme. *Water Res.*, 40, 1686–1696.

Karvelas, M., Katsoyiannis, A., Samara, C. (2003). Occurrence and fate of heavy metals in the wastewater treatment process. *Chemosphere*, 53, 1201–1210.

Kinney, C.A., Furlong, E.T., Zaugg, S.D., Burkhardt, M.R., Werner, S.L., Cahill, J.D., Jorgensen, G.R. (2006). Survey of organic wastewater contaminants in biosolids destined for land application. *Environ. Sci. Technol.*, 40(23), 7207–7215.

Kreuzinger N. (ed.) (2002). Occurrence of highly discussed pollutants in the stretch of the Austrian Danube related to the Catchment Area. In *Oral Presentation at SETAC Europe 12th Annual Meeting*, Vienna.

Kreuzinger, N., Clara, M., Strenn, B., Kroiss, H. (2004). Relevance of the sludge retention time (SRT) as design criteria for wastewater treatment plants for the removal of endocrine disruptors and pharmaceuticals from wastewater. *Water Sci. Technol.*, 50 (5), 149–156.

Kümmerer, K. (2001). Drugs in the environment: Emission of drugs, diagnostic aids and disin-fectants into wastewater by hospitals in relation to other sources – A review. *Chemosphere*, 45, 957–969.

Kümmerer, K. (2009). The presence of pharmaceuticals in the environment due to human use: Present knowledge and futurs challenges. *J. Environ. Manage.*, 90, 2354–2366.

Lassachagne, D. (2014). Devenir de micropolluants présents dans les boues d'épuration, du traitement a l'épandage agricole : application aux micropolluants métalliques (Cd, Cu) et organiques (médicaments) issus du traitement biologique conventionnel d'effluents urbains ou hospitaliers. PhD Thesis, University of Limoges.

Lindberg, R.H., Jarnheimer, P.-Å., Olsen, B., Johansson, M., Tysklind, M. (2004). Determination of antibiotic substances in hospital sewage water using solid phase extraction and liquid chromatography/mass spectrometry and group analogue internal standards. *Chemosphere*, 57, 1479–1488.

Lindberg, R.H., Wennberg, P., Johansson, M.I., Tysklind, M., Andersson, B.A.V. (2005). Screening of human antibiotic substances and determination of weekly mass flows in five sewage treatment plants in Sweden. *Environ. Sci. Technol.*, 39, 3421–3429.

Lindberg, R.H., Björklund, K., Rendahl, P., Johansson, M. J., Tysklind. M., Andersson, B.A.V. (2006). Environmental risk assessment of antibiotics in the Swedish environment with emphasis on sewage treatment plants. *Water Res.*, 41(3), 613–619.

MacKay, A.A. and Vasudevan, D. (2012). Polyfunctional ionogenic compound sorption: Challenges and new approaches to advance predictive models. *Environ. Sci. Technol.*, 46(17), 9209–9223.

Magdeburg, A., Stalter, D., Schlüsener, M., Ternes, T., Oehlmann, J. (2014). Evaluating the efficiency of advanced wastewater treatment: Target analysis of organic contaminants and (geno-)toxicity assessment tell a different story. *Water Res.*, 50, 35–47.

Majewsky, M., Galle, T., Yargeau, V., Fischer, K. (2011). Active heterotrophic biomass and slu-dge retention time (SRT) as determining factors for biodegradation kinetics of pharmaceuticals in activated sludge. *Bioresour. Technol.*, 102, 7415–7421.

Martín, J., Camacho-Muñoz, D., Santos, J.L., Aparicio, I., Alonso, E. (2012). Occurrence of pharmaceutical compounds in wastewater and sludge from wastewater treatment plants: Removal and ecotoxicological impact of wastewater discharges and sludge disposal. *J. Hazard. Mater.*, 239–240, 40–47.

Metcalf & Eddy (2004). *Wastewater Engineering: Treatment and Reuse*. McGraw-Hill, Boston.

Miao, X.-S., Koenig, B.G., Metcalfe, C.D. (2002). Analysis of acidic drugs in the effluents of sewage treatment plants using liquid chromatography–electrospray ionization tandem mass spectrometry. *J. Chromatogr.*, A 952, 139–147.

Miège, C., Choubert, J.M., Ribeiro, L., Eusèbe, M., Coquery, M. (2009). Fate of pharmaceuticals and personal care products in wastewater treatment plants – Conception of a database and first results. *Environ. Pollut.*, 157, 1721–1726.

Mullot, J.-U. (2009). Modélisation des flux de médicaments dans les effluents hospitaliers. PhD Thesis, Paris-Sud University, Hauts-de-Seine.

Mullot, J.-U., Karolak, S., Fontova, A., Levi, Y. (2010). Modeling of hospital wastewater pollution by pharmaceuticals: First results of Mediflux study carried out in three French hospitals. *Water Sci. Technol.*, 62, 2912–2919.

Nielsen, P.H., Thomsen, T.R., Nielsen, J.L. (2004). Bacterial composition of activated sludge – Importance for floc and sludge properties. *Water Sci. Technol.*, 49, 51–58.

Nielsen, P.H., Mielczarek, A.T., Kragelund, C., Nielsen, J.L., Saunders, A.M., Kong, Y.H., Hansen, A.A., Vollertsen, J. (2010). A conceptual ecosystem model of microbial communities in enhanced biological phosphorus removal plants. *Water Res.*, 44(17), 5070–5088.

Nsenga Kumwimba, M. and Meng, F. (2019.), Roles of ammonia-oxidizing bacteria in improving metabolism and co-metabolism of trace organic chemicals in biological wastewater treatment process: A review. *Sci. Total Environ.*, 659, 419–441.

Oberoi, A.S., Jia, Y., Zhang, H., Khanal, S.K., Lu, H. (2019). Insights into the fate and removal of antibiotics in engineered biological treatment systems: A critical review. *Environ. Sci. Technol.*, 53, 7234–7264.

Pagnanelli, F., Moscardini, E., Giuliano, V., Toro, L. (2004). Sequential extraction of heavy metals in river sediments of an abandoned pyrite mining area: Pollution detection and affinity series. *Environ. Pollut.*, 132, 189–201.

Pagnanelli, F., Mainelli, S., Bornoroni, L., Dionisi, D., Toro, L. (2009). Mechanisms of heavy metal removal by activated sludge. *Chemosphere*, 75, 1028–1034.

Pan, B., Zhang, W., Lv, L., Zhang, Q., Zheng, S. (2008). Development of polymeric and polymer-based hybrid adsorbents for pollutants removal from waters. *Chem. Eng. J.*, 151, 19–29.

Pan, B., Huang, P., Wu, M., Wang, Z., Wang, P., Jiao, X., Xing, B. (2012). Physicochemical and sorption properties of thermally-treated sediments with high organic matter content. *Bioresour. Technol.*, 103, 367–373.

Park, J., Yamashita, N., Wu, G., Tanaka, H. (2017). Removal of pharmaceuticals and personal care products by ammonia oxidizing bacteria acclimated in a membrane bioreactor: Contributions of co-metabolism and endogenous respiration. *Sci. Total Environ.*, 605–606, 18–25.

Petrović, M., Gonzalez, S., Barceló, D. (2003). Analysis and removal of emerging contaminants in wastewater and drinking water. *Trends Anal. Chem.*, 22, 685–696.

Pomiès, M. (2013). Etude et modélisation dynamique de l'élimination de micropolluants prioritaires et émergents au sein du procédé à boues activées, PhD Thesis, University of Montpellier 1, Montpellier.

Pomiès, M., Choubert, J.-M., Wisniewski, C., Coquery, M. (2013). Modelling of micropollutant removal in biological wastewater treatments: A review. *Sci. Total Environ.*, 443, 733–748.

Rivera-Utrilla, J., Sánchez-Polo, M., Ferro-García, M.Á., Prados-Joya, G., Ocampo-Pérez, R. (2013). Pharmaceuticals as emerging contaminants and their removal from water: A review. *Chemosphere*, 93(7), 1268–1287.

Rogers, H.R. (1996). Sources, behaviour and fate of organic contaminants during sewage treatment and in sewage sludge. *Sci. Total Environ.*, 185, 3–26.

Rybacka, A. and Andersson, P.L. (2016). Considering ionic state in modeling sorption of pharmaceuticals to sewage sludge. *Chemosphere*, 165, 284–293.

Schmidt, S., Winter, J., Gallert, C. (2012). Long-term effects of antibiotics on the elimination of chemical oxygen demand, nitrification, and viable bacteria in laboratory-scale wastewater treatment plants. *Arch. Environ. Contam. Toxicol.*, 63, 354–364.

Schwarzenbach, R.P., Gschwend, P.M., Imboden, D.M. (2003). *Environmental Organic Chemistry*, 2nd edition. John Wiley & Sons, New York.

Seira, J. (2013). Role de la sorption et de la biodégradation dans l'élimination de micropolluants par des procédés d'épuration biologique : application aux molécules anticancéreuses traitées par bioréacteur à membrane. INP ENSIACET, University of Toulouse, Toulouse.

Soulier, C., Gabet, V., Lardy, S., Lemenach, K., Pardon, P., Esperanza, M., Miège, C., Choubert, J.-M., Martin, S., Bruchet, A. et al. (2011). Zoom on pharmaceutical substances: Presence, distribution and behaviour in wastewater treatment plants. *Tech. Sci. Méthodes*, 1/2, 63–77.

Stadler, L.B. and Love, N.G. (2016). Impact of microbial physiology and microbial community structure on pharmaceutical fate driven by dissolved oxygen concentration in nitrifying bio-reactors. *Water Res.*, 104, 189–199.

Stadler, L.B., Delgado Vela, J., Jain, S., Dick, G.J., Love, N.G. (2018). Elucidating the impact of microbial community biodiversity on pharmaceutical biotransformation during wastewater. *Treatment, Microb. Biotechnol.*, 11, 995–1007.

Stadlmair, L.F., Letzel, T., Drewes, J.E., Grassmann, J. (2018). Enzymes in removal of pharmaceuticals from wastewater: A critical review of challenges, applications and screening methods for their selection. *Chemosphere*, 205, 649–661.

Stalder, T. (2012). Implication des effluents d'activités hospitalières et de la filière carnée sur la dissémination de l'antibiorésistance : dynamique des intégrons de l'émission au rejet. PhD Thesis, University of Limoges.

Stalder, T., Barraud, O., Casellas, M., Dagot, C., Ploy, M.-C. (2012). Integron involvement in environmental spread of antibiotic resistance. *Antimicrob. Resist. Chemother.*, 3, 119.

Suarez, S., Lema, J.M., Omil, F. (2010). Removal of pharmaceutical and personal care products (PPCPs) under nitrifying and denitrifying conditions. *Water Res.*, 44, 3214–3224.

Suarez, S., Omil, F., Lema, J.M. (2019). Pre-treatment of hospital wastewater by coagulation–flocculation and flotation. *Bioresour. Technol.*, 100, 2138–2146.

Tadkaew, N., Hai, F.I., McDonald, J.A., Khan, S.J., Nghiem, L.D. (2011). Removal of trace organics by MBR treatment: The role of molecular properties. *Water Res.*, 45, 2439–2451.

Tahar, A. (2013). Amélioration de l'élimination des micropolluants organiques des eaux usées par traitements secondaire et tertiaire. PhD Thesis, University of Montpellier 1, Montpellier.

Tolls, J. (2001). Sorption of veterinary pharmaceuticals in soils: A review. *Environ. Sci. Technol.*, 35, 3397–3406.

Torresi, E., Fowler, S.J., Polesel, F., Bester, K., Andersen, H.R., Smets, B.F., Plósz, B.G., Christensson, M. (2016). Biofilm thickness influences biodiversity in nitrifying MBBRs – Implications on micropollutant removal. *Environ. Sci. Technol.*, 50, 9279–9288.

Torresi, E., Polesel, F., Bester, K., Christensson, M., Smets, B.F., Trapp, S., Andersen, H.R., Plósz, B.G. (2017). Diffusion and sorption of organic micropollutants in biofilms with varying thickness. *Water Res.*, 123, 388–400.

Tran, H.N., Urase, T., Ngo, H.H., Hu, J., Ong, S.L. (2013). Insight into metabolic and co-metabolic activities of autotrophic and heterotrophic microorganisms in the biodegradation of emerging trace organic contaminants. *Bioresour. Technol.*, 146, 721–731.

Trapp, S., Franco, A., MacKay, D. (2010). Activity-based concept for transport and partitioning of ionizing organics. *Environ. Sci. Technol.*, 44, 6123–6129.

Verlicchi, P., Al Aukidy, M., Galletti, A., Petrovic, M., Barceló, D. (2012a). Hospital effluent: Investigation of the concentrations and distribution of pharmaceuticals and environmental risk assessment. *Sci. Total Environ.*, 30, 109–118.

Verlicchi, P., Al Aukidy, M., Zambello, E. (2012b). Occurrence of pharmaceutical compounds in urban wastewater: Removal, mass load and environmental risk after a secondary treatment: A review. *Sci. Total Environ.*, 429, 123–155.

Vieno, N.M., Tuhkanen, T., Kronberg, L. (2005). Seasonal variation in the occurrence of pharmaceuticals in effluents from a sewage treatment plant and in the recipient water. *Environ. Sci. Technol.*, 39(21), 8220–8226.

Vieno, N.M., Tuhkanen, T., Kronberg, L. (2007). Elimination of pharmaceuticals in sewage treatment plants in Finland. *Water Res.*, 41, 1001–1012.

Wolff, D., Krah, D., Dötsch, A., Ghattas, A.K., Wick, A., Ternes, T.A. (2018). Insights into the variability of microbial community composition and micropollutant degradation in diverse biological wastewater treatment systems. *Water Res.*, 143, 313–324.

Wolff, D., Helmholz, L., Castronovo, S., Ghattas, A.K., Ternes, T.A., Wick, A. (2021). Micro-pollutant transformation and taxonomic composition in hybrid MBBR – A comparison of carrier-attached biofilm and suspended sludge. *Water Res.*, 202, 117441.

Wunder, D.B., Bosscher, V.A., Cok, R.C., Hozalski, R.M. (2011). Sorption of antibiotics to biofilm. *Water Res.*, 45, 2270–2280.

Xu, K., Harper, W.F., Zhao, D. (2008). 17α-Ethinylestradiol sorption to activated sludge bio-mass: Thermodynamic properties and reaction mechanisms. *Water Res.*, 42, 3146–3152.

Xu, J., Sheng, G.P., Ma, Y., Wang, L.F., Yu, H.Q. (2013). Roles of extracellular polymeric substances (EPS) in the migration and removal of sulfamethazine in activated sludge system. *Water Res.*, 47, 5298–5306.

Yu, Y. and Wu, L. (2012). Analysis of endocrine disrupting compounds, pharmaceuticals and personal care products in sewage sludge by gas chromatography–mass spectrometry. *Talanta*, 89, 258–263.

Yu, T.H., Lin, A.Y.C., Lateef, S.K., Lin, C.F., Yang, P.Y. (2006). Removal of antibiotics and non-steroidal anti-inflammatory drugs by extended sludge age biological process. *Chemosphere*, 77, 175–181.

Yu, Z., Yediler, A., Yang, M., Schulte-Hostede, S. (2012). Leaching behaviour of enrofloxacin in three different soils and the influence of a surfactant on its mobility. *J. Environ. Sci.*, 24, 435–439.

Yuncu, B., Sanin, F.D., Yetis, U. (2006). An investigation of heavy metal biosorption in relation to C/N ratio of activated sludge. *J. Hazard. Mater.*, 137, 990–997.

Zeng, Q., Li, Y., Gu, G. (2009). Effect of temperature on the sorption of 17α-ethinylestradiol to aerobic and anaerobic sludge. In *2009 International Conference on Energy and Environment Technology*, 294–297.

Zhang, H., Jia, Y., Khanal, S.K., Lu, H., Fang, H., Zhao, Q. (2018). Understanding the role of extracellular polymeric substances on ciprofloxacin adsorption in aerobic sludge, anaerobic sludge, and sulfate-reducing bacteria sludge systems. *Environ. Sci. Technol.*, 52, 6476–6486.

2

Lagoons and Oxidation Lagoons

The simplest way to purify wastewater is to discharge it into ponds and allow natural forces such as sunlight, bacteria, algae, temperature and time to take their course.

This system goes by different names depending on the author: lagoon, pond, oxidation pond, stabilization pond.

Lagoons are one of the most common wastewater treatment options because they are cost-effective and easy to maintain. They are relatively inexpensive in terms of equipment, maintenance, operating costs, energy costs and labor. When wastewater enters a large-capacity lagoon, it remains there for an extended period. This allows bacteria to multiply and break down many components of the wastewater.

However, the solids present in wastewater and the bacteria that develop on the contaminants tend to precipitate or accumulate. Therefore, they must be dredged or removed periodically.

This technique uses only extremely simple technology, requiring minimal industrial-type structures and equipment, and its operation consumes little or no energy.

A stabilization lagoon or pond is therefore defined as any natural or artificial depression or excavation into which raw or decanted wastewater flows naturally, emerging without any external intervention in a state where it does not risk altering the quality of the host environment.

2.1. Description

A lagoon is part of a natural wastewater treatment method carried out using aquatic vegetation and microorganisms, primarily algae and bacteria. This system is used both inland and along the coastline.

A wastewater treatment plant is primarily a series of three to five ponds, ranging from 40 cm to 120 cm in depth. These ponds are often built within wetlands, where water flows under the force of gravity. The first ponds are microorganism ponds, where the organic matter in the wastewater breaks down.

A food chain then develops: phytoplankton, zooplankton, daphnia, etc. Under the influence of sunlight, ultraviolet rays destroy some of the bacteria. The water then flows into shallower ponds with plant life similar to that found along the edges of natural lakes, streams, backwaters, dams, reservoirs and other bodies of water: irises, reeds, rushes, etc. These plants absorb the minerals produced by the decomposition of organic matter for their growth.

Lagoons are generally classified into three categories:

– aerobic lagoons;

– anaerobic lagoons;

– optional lagoons.

Figure 2.1. *Wastewater pretreatment*

2.1.1. *Aerated lagoons*

Aerated lagoons are primarily an aerobic biological treatment process using free cultures, which differs from activated sludge due to the absence of sludge recirculation. The bacterial population is therefore in equilibrium with the carbon substrate, which has two consequences:

– The bacterial density remains modest, requiring a long treatment time to obtain a small amount of dissolved residue.

– Flocculation is relatively weak, which demands the use of a large settling lagoon as the site of slow separation.

A treatment plant with aerated lagoons typically comprises two ponds:

– an aerated lagoon equipped with surface aerators;

– a settling lagoon.

The aerated lagoon should have a depth of at least 2.5–3 m (depth increasing with the unit power of the aerators). The settling lagoon should have a typical depth of 2 m to leave 1 m of free water before the sludge is removed.

Depending on the size of the station, there may be a single settling lagoon with a retention time of five days, or two lagoons operating alternately and having a shorter retention time (four days).

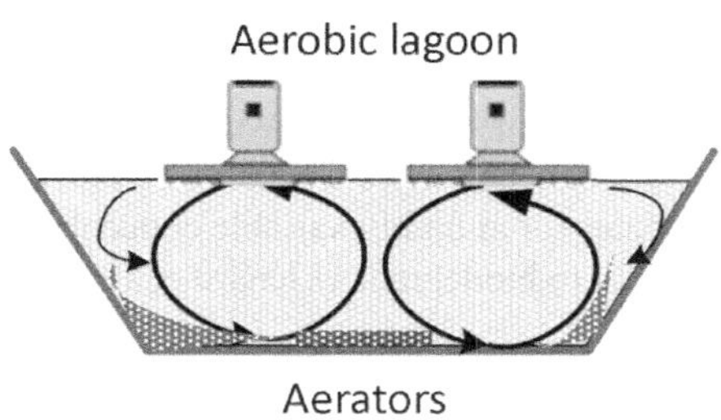

Figure 2.2. *Operating principle of an aerobic lagoon*

Approximately every two years, sludge extraction is necessary, which is facilitated by a configuration with two settling lagoons (Alexandre et al. 1997).

Aerobic lagoons have surface aerators that ensure homogenization of the environment and also increase dissolved oxygen. These lagoons are rarely completely mixed or maintain a significant amount of dissolved oxygen throughout the entire water volume, so anaerobic processes occur in the bottom sediments and in "dead zones" that are not completely mixed or aerated.

Aerated lagoons are very effective at removing biodegradable BOD_5 and nitrification. Aerobic bacteria produce more biomass than anaerobic bacteria for an equivalent concentration of organic matter removed, expressed as BOD_5. Therefore, more biomass is produced. This accumulates at the bottom of the lagoons or can be recovered in clarifiers that can be placed downstream.

If a clarifier is used, the bacteria can be recycled to the aerated zone and the lagoon will function as an activated sludge system with a higher bacterial concentration, thus requiring a smaller lagoon to meet effluent targets.

Problems that frequently occur in aerated lagoons include:

– Low BOD_5 removal: the absence of essential nutrients (often trace elements) or inhibition by chemicals in wastewater can slow or stop bacterial growth. In rare cases, bacteria may settle too quickly or remain suspended and be removed with the effluent, leaving a concentration that is too low in the water column.

– Foams: caused by surfactants present in wastewater or by filamentous bacteria that develop due to specific conditions such as low dissolved oxygen levels, unbalanced nutrients, high fat, oil, and grease content, etc.

– Dispersed biomass: high levels of suspended solids in the effluent can be due to high salt concentrations, toxic compounds or rapid bacterial growth when large masses of highly concentrated organic matter enter the lagoon.

A type of bacteria necessary for the breakdown of a chemical compound may be absent or present in too low a concentration. This may be due to the sporadic presence of the compound in wastewater (or at highly variable concentrations) or to the fact that the compound may be difficult to biodegrade (Liénard et al. 2004).

Aerobic ponds are characterized by the fact that only aerobic microorganisms are active. Bacteria and algae degrade organic matter together with physical factors such as wind and light.

In these ponds, algae produce large volumes of oxygen through photosynthesis, allowing the environment to remain aerobic.

The air–water exchange surface also promotes the dissolution of oxygen in the water, helping to increase its concentration in the environment.

However, aerated lagoons have both advantages and disadvantages. The advantages are as follows:

– a rather rustic method;

– process adapted to significant variations in hydraulic or organic loads;

– good tolerance to highly concentrated effluents, nutrient-unbalanced effluents or even the occasional arrival of toxic products.

The disadvantages are as follows:

– average quality discharges across all parameters;

– presence of some electromechanical equipment requiring specific maintenance;

– energy-intensive process (compared to natural lagoons);

– sensitivity to poor upstream screening (Besnault and Prost-Boucle 2004);

– low expected performance:

BOD_5: 35 $mg.L^{-1}$

– elimination of approximately 25–30% nitrogen and phosphorus: (very little elimination beyond assimilation by bacteria).

2.1.2. *Anaerobic lagoons*

If the wastewater is not aerated, the lagoon can operate anaerobically, where organic matter is fermented into simple organic acids and ultimately converted into methane. Anaerobic lagoons rely on bacterial growth to remove contaminants. However, because anaerobic bacteria can grow more slowly than aerobic bacteria and can settle at the bottom if not mechanically mixed, the bacterial concentration is much lower than in an activated sludge process.

The removal of contaminants measured in BOD_5 or COD is often in the range of 50–70%. As a general rule, anaerobic lagoons are the first stage of a multi-stage treatment process.

Problems that frequently occur in anaerobic lagoons include:

– An acidic environment: fermentation reactions produce simple acids that lower the pH. Methanogens, which normally metabolize acids and neutralize the pH, may not grow properly, and a low pH inhibits the growth of methanogens and other types of bacteria.

– Low BOD_5 removal: the absence of essential nutrients (often trace elements) or inhibition by chemicals in wastewater can slow or stop bacterial growth. In rare cases, bacteria may settle too quickly or remain suspended and be removed with the effluent, leaving a concentration that is too low in the water column.

– Low methane production: similar to low BOD_5 removal, but may be due to limited nutrients specifically required by methanogens or to inhibitors specific to them.

– Strong odor production: anaerobic processes such as sulfate reduction generate hydrogen sulfide. The simple acids produced are volatile and include acetic, propionic and butyric acids, all of which produce bad odors. Ammonia is also released and, depending on the pH and concentration, can contribute to an unpleasant smell.

Anaerobic lagoons are characterized by an absence of oxygen in the environment. The decomposition of organic matter is carried out by anaerobic bacteria. The depth of the ponds can reach 4–5 m.

The accumulation of sludge at the bottom of the ponds promotes digestion, which results in the production and release of gas bubbles towards the surface of the pond. The dispersion of solid particles bound to these gas bubbles tends to create areas of water turbidity, inhibiting the growth of algae.

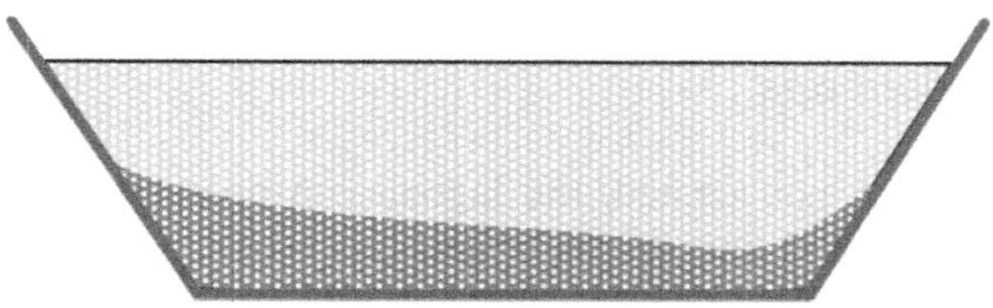

Figure 2.3. *Anaerobic lagoon*

Anaerobic lagoons are used as the primary treatment process. The lagoons are oxygen-free and designed to reduce BOD and solid concentrations in wastewater through sedimentation and anaerobic digestion. They are particularly useful for reducing high BOD and TSS concentrations from agricultural and food processing wastewater. Anaerobic digestion occurs mainly in the sludge at the bottom of the lagoon, converting organic matter into methane and carbon dioxide and releasing certain soluble byproducts into the water column (e.g. organic acids, ammonia). At temperatures below 12°C, digestion processes slow down. Anaerobic lagoons are generally more than 2 m deep and dedicated solely to sludge storage. The total height can reach over 3.5 m. The hydraulic retention time depends on the required BOD volumetric loading rate ($g.m^{-3}.d^{-1}$) for the climate and can last up to 20 days. The organic matter degradation (BOD) is between 40 and 70%. This efficiency, combined with the partial mineralization of organic matter, allows for the subsequent use of smaller aerobic ponds. The main disadvantages of anaerobic lagoons are the risks of odors and increased ammonia and gas levels (Pearson et al. 1987).

2.1.3. *Facultative lagoons*

Facultative lagoons can be actively aerated or simply use oxygen from the air that diffuses into the surface water. Ideally, aerobic and anaerobic bacteria will be present and active and contribute to the removal of contaminants.

Often, a lower layer of anaerobic water and sediments provides the anaerobic environment for the same processes as the anaerobic lagoon, but aerobic bacteria present in the upper water layers can perform additional metabolic processes such as nitrification and more efficient removal of the BOD_5. Volatile acids are converted into carbon dioxide and water, and hydrogen sulfide can be oxidized into elemental sulfur or sulfate compounds.

Nitrites and nitrates formed in the aerobic zone can be reduced to nitrogen gas in the anaerobic zone, thus removing some of the dissolved nitrogen. With sufficient retention time, wastewater can be clarified with a 90% BOD_5 removal rate.

Problems that occur in facultative lagoons include those encountered in anaerobic lagoons, plus:

– Foams: these can be caused by surfactants present in wastewater or by filamentous bacteria that develop due to specific conditions such as low dissolved oxygen levels, unbalanced nutrients, high fat, oil, and grease content, etc.

– Dispersed biomass: high levels of suspended solids in the effluent can be due to high salt concentrations, toxic compounds or rapid bacterial growth when large masses of highly concentrated organic matter enter the lagoon.

– Water layer inversion: variations in wind and/or temperature can cause the surface and lower layers of the lagoon to mix. Sediments and odors from the bottom mix with the aerated water above, thus disrupting aerobic and anaerobic processes.

– Nitrification: nitrifying bacteria are sensitive to toxins, low pH and mineral deficiencies. Because they grow slowly, they can be eliminated from the system if the flow rate increases or if their growth is slowed by the conditions mentioned above.

Facultative lagoons result from a combination of the two previous lagoon types. The depth of the lagoons is approximately 1–2.5 meters. Two zones are thus observed: an aerobic zone in the upper part of the lagoon, and an anaerobic zone in the lower part.

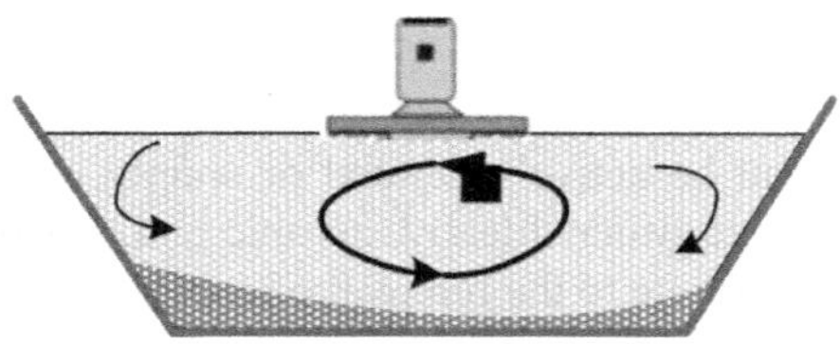

Figure 2.4. *The operation of a facultative lagoon*

These lagoons can be used for tertiary treatment to remove pathogens, nutrients and potentially algae. They are very shallow (generally about 1 m deep) to allow light to penetrate to the bottom and maintain aerobic conditions throughout. They are installed downstream of aerated lagoons. Their size and number are determined by the retention time required to reach a specific concentration of pathogens in the effluent. In the absence of pathogen effluent limits, these structures act as buffers in case of optional failure of the upstream lagoons.

Facultative lagoons perform primary, secondary and tertiary treatments. Lagoons used for the primary treatment receive raw wastewater, generally after mechanical pretreatment, and then store suspended solids as sludge. Facultative lagoons used for the secondary treatment follow the primary treatment process; for example, an anaerobic lagoon. In this case, these lagoons are not supplied with the same pollutant loads.

The main characteristics of these ponds are the presence of a permanent anaerobic layer with sludge accumulation. Therefore, a deep zone can be provided at the lagoon inlet for the deposition of heavier materials that settle easily. An aerobic zone develops at the surface, oxygenated by the photosynthetic mechanisms of algae and the turbulence created by the wind. Depending on the oxygen consumption rate of the environment, the aerobic layer can become anoxic at night when photosynthesis ceases.

Facultative lagoons are generally shallow (1–3 m deep) to allow adequate light penetration while having sufficient depth to prevent the emergence of macrophytes.

Primary facultative lagoons are generally rectangular, with a length-to-width ratio of 2–5:1 to prevent excessive sludge accumulation around the inlet. The ponds are regularly shaped and do not contain any isolated areas where hydraulic circulation is impaired and that are conducive to weed growth.

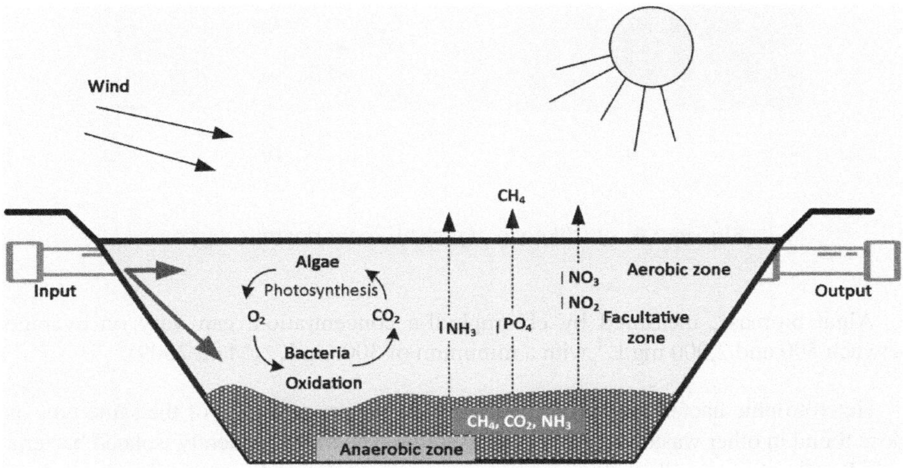

Figure 2.5. *Facultative lagoon*

The biomass in facultative primary lagoons consists mainly of microalgae, facultative bacteria and anaerobic bacteria, although a wide variety of other organisms, such as protozoa, rotifers, fly larvae, viruses and fungi, may also be present depending on the time of year.

A representative diagram of the major interactions in facultative lagoons is shown in Figure 2.5 (Metcalf & Eddy 1991).

Algae occupy the surface layer where light can penetrate (the photic zone) and give facultative lagoons their characteristic green color. The aerobic zone is generally 40–60 cm deep and includes motile genera such as *Chlamydomonas*, *Euglena* and *Phacus,* which tend to dominate (Mara and Pescod 1998). This is due to their ability to adjust their vertical position to optimize access to light in the pond's murky water. The motile algae population tends to form in a narrow band approximately 20 cm thick, which rises and falls in the water column during the day and disperses at night (Mara 2009). Non-motile algae types such as *Chlorella* and *Scenedesmus* are also abundant.

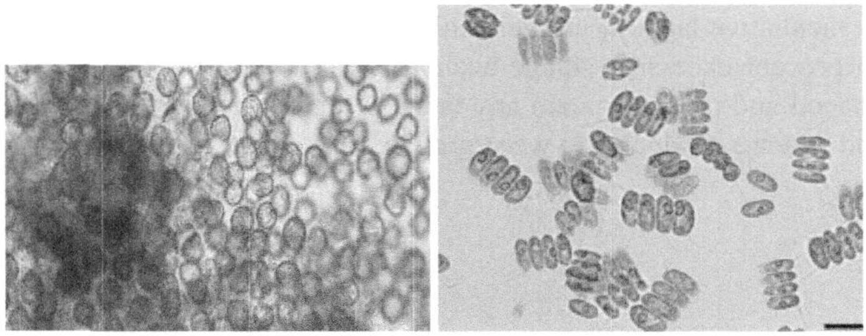

a) b)

Figure 2.6. *a) Chlorella algae. b) Scenedesmus algae*

Algal biomass, measured by chlorophyll-a concentration, can vary on average between 500 and 2,000 mg.L^{-1} with a minimum of 300 mg.L^{-1} (Mara 2009).

Heterotrophic bacteria present in the aerobic zone are generally of the same type as those found in other wastewater treatment systems. The most frequently isolated bacteria include *Beggiatoa alba*, *Sphaerotilus natans*, *Achromobacter*, *Flavobacterium*, *Pseudomonas* and *Zoogloea* spp. (EPA 2011). In temperate climates, mesophilic and psychrophilic bacteria dominate in summer, and psychrophilic and psychotropic bacteria dominate in winter (Townshend and Knoll 1987).

Below the algal zone, where dissolved oxygen is low or even nonexistent and where only long wavelengths (750–900 nm) penetrate (chlorophyll-a absorbs at 663 nm), photosynthetic anaerobic bacteria such as *Chromatiaceae* (purple sulfur bacteria; retain S as granules) and *Chlorobiaceae* (green sulfur bacteria; release S) proliferate. These sulfur bacteria use H_2S for photosynthesis, producing sulfur compounds as byproducts. In the presence of light, they can also oxidize sulfides to sulfates (Pearson et al. 1996).

At the bottom of the lagoon lies the anaerobic mud layer. A wide range of proteolytic bacteria such as *Streptococcus*, *Clostridium* and *Staphylococcus* break down proteins and amino acids into simple organic acids, ammonia, carbon dioxide, hydrogen and hydrogen sulfide. Acetogenic bacteria are obligately anaerobic and use hydrogen and carbon dioxide to produce acetate. Methanogenic bacteria such as *Methanobacterium* and *Methanospirillum* convert acetate and hydrogen into methane. Methanogenesis can remove significant amounts of BOD from sludge but requires a narrow pH range (6.4–7.2) and moderate to high temperatures (above 15°C). Sulfate-reducing bacteria (e.g. desulfovibrio) also inhabit the sediments, converting sulfates into sulfides. These can then be oxidized to sulfur or sulfates, either by purple bacteria in the facultative zone or by aerobic bacteria such as *Thiobacillus* through direct combination with dissolved oxygen.

In the liquid column of a facultative pond, aerobic and anaerobic conditions can alternate. The rate at which the pond water oscillates between aerobic and anaerobic conditions depends primarily on the balance between respiration and photosynthesis rates, which in turn depend on temperature, light availability, the relative number of algae and bacteria present and nutrient availability. Anaerobic conditions tend to occur: 1) during the night when algae respire; 2) in calm, warm weather, when oxygen solubility in water is low and wind mixing is insufficient; or 3) continuously if the pond is overloaded. Aerobic conditions generally occur: 1) when morning sun activates photosynthesis; 2) when windy conditions aerate the surface or mix stratified layers; and 3) when the load is reduced according to climatic conditions (Zouhir 2008).

Figure 2.7. *Lagoon invaded by algae*

Since algae are the primary source of dissolved oxygen, the concentration reaches a maximum when the intensity of incident sunlight is at its highest. In the presence of an abundant algal population, the relative rate of photosynthesis can generally lead to supersaturation of dissolved oxygen concentrations during the day. Under calm conditions, the formation of an algal bloom obscures the lower areas, and the immobile algae emerge from the photic zone and begin to respire. In cold and temperate weather, the algae may not develop year-round, and therefore the oxygen supply may be unreliable.

2.2. Design

The use of lagoons varies depending on the desired treatment, and some authors refer to:

– a complete lagoon when the installation is directly supplied with raw water;

– a secondary lagoon when the installation is supplied with decanted water;

– a tertiary lagoon for a facility supplied with effluent clarified using a conventional process (activated sludge, biofilters, MBBR).

2.2.1. *Components of a lagoon system*

A typical lagoon system consists of one or more shallow lagoons, designed to retain wastewater for a specific period. These lagoons are generally constructed with earthen berms or embankments and lined with clay or synthetic materials to prevent wastewater leakage into the surrounding soil and groundwater.

The number and configuration of lagoons in a system can vary depending on specific treatment objectives, local climate and available land area. However, most lagoon systems include at least one primary lagoon for initial treatment and a secondary lagoon for further treatment and stabilization:

– The primary lagoon, also called the "anaerobic lagoon", is the first stage of treatment in a lagoon system. It is designed to receive and retain raw wastewater, allowing solids to settle and anaerobic bacteria to begin breaking down organic matter. The anaerobic conditions in the primary lagoon promote the growth of bacteria that thrive in the absence of oxygen. These bacteria transform organic matter into gases such as methane and carbon dioxide, which are released into the atmosphere. This process also helps reduce odors and pathogens in wastewater.

– The secondary lagoon, also called the aerobic lagoon, is the second stage of treatment in a lagoon system. It is designed to further treat and stabilize wastewater after it has passed through the primary lagoon. The aerobic conditions of the secondary lagoon promote the growth of bacteria that thrive in the presence of oxygen. These bacteria continue the decomposition of organic matter, transforming it into byproducts such as water and carbon dioxide. The secondary lagoon also provides a habitat for algae, which produce oxygen through photosynthesis and further purify the wastewater.

2.2.1.1. *Processes involved in a lagoon system*

The lagoon system relies on a series of natural processes to treat wastewater. These processes can be classified as physical, biological and chemical, each playing a crucial role in the overall treatment process.

Physical processes include sedimentation, evaporation and disinfection by sunlight. Biological processes involve the decomposition of organic matter by bacteria and other microorganisms. Chemical processes include the neutralization of harmful substances and the conversion of nutrients into less harmful forms.

2.2.1.1.1. Physical processes

Sedimentation is a key physical process in a lagoon. As wastewater enters the primary lagoon, the water velocity slows, allowing solids to settle to the bottom. Over time, these solids form a sludge layer, which is periodically removed for disposal or further treatment.

Evaporation and disinfection by sunlight are also important physical processes. The shallow depth and large surface area of the lagoons promote evaporation, which helps reduce the volume of wastewater. Sunlight, on the other hand, can kill or inactivate pathogens present in the wastewater, thus contributing to the disinfection process.

2.2.1.1.2. Biological processes

Biological processes are central to a lagoon system. These processes involve the decomposition of organic matter by bacteria and other microorganisms. Under the anaerobic conditions of the primary lagoon, bacteria transform organic matter into gases such as methane and carbon dioxide. Under the aerobic conditions of the secondary lagoon, bacteria and algae continue this decomposition process, transforming organic matter into harmless byproducts such as water and carbon dioxide.

The balance between bacteria and algae in a lagoon system is crucial for effective treatment. An excess of bacteria can deplete the water's oxygen, leading to odors and poor treatment performance. Conversely, an excess of algae can cause problems such as algal blooms and oxygen depletion at night when photosynthesis ceases.

2.2.1.1.3. Chemical processes

Chemical processes in a lagoon system include the neutralization of harmful substances and the conversion of nutrients into less harmful forms. For example, ammonia, a harmful nutrient found in wastewater, can be converted into nitrate, a less harmful form, through a process called nitrification. This process is carried out by specific bacteria in the presence of oxygen.

Another important chemical process is phosphorus removal. Phosphorus, another nutrient present in wastewater, can be removed by precipitation, adsorption or biological absorption. However, the efficiency of these processes can vary depending on the specific conditions of the lagoon.

2.2.1.2. *Advantages and disadvantages of lagoons*

2.2.1.2.1. Advantages

The simplicity of a lagoon system is one of its main advantages. Its operation requires minimal mechanical equipment and technical expertise, making it suitable for rural areas and small communities. Low maintenance requirements also mean it is less likely to break down or require costly repairs.

Another advantage of lagoons is their low energy consumption. Unlike mechanical treatment methods, which require electricity to run pumps and other equipment, a lagoon system relies on natural processes to treat wastewater. This not only reduces operating costs but also makes it more environmentally friendly.

2.2.1.2.2. Disadvantages

One of the main drawbacks of lagoons is their land requirement. Lagoons need a large area of land to operate effectively, which can be a limitation in densely populated areas or areas where land costs are high. This can also limit the system's scalability, making it less suitable for large communities or cities.

Another drawback is sludge production. Over time, solids settle at the bottom of the lagoons, forming a layer of sludge. This sludge must be periodically dredged and disposed of, which can be a costly and difficult process. Furthermore, the sludge can contain harmful substances, posing an environmental risk if not properly managed.

In conclusion, a lagoon system is a natural and economical method of wastewater treatment. It relies on a series of physical, biological and chemical processes to purify wastewater, making it suitable for discharge or reuse. Despite its limitations, it remains a popular choice for wastewater treatment in rural areas and small communities.

2.2.2. *Lagoon aeration*

An aerated lagoon can be one of the most efficient, cost-effective and low-maintenance wastewater treatment methods. Since aeration can account for 50–60% of energy costs, it is important to consider design conditions when determining the type and amount of aeration required.

Figure 2.8. *Aerobic lagoons*

2.2.3. *Performance*

2.2.3.1. *Removal of BOD_5*

A lagoon's capacity to remove BOD_5 is the primary factor by which its performance is evaluated. BOD_5 degradation efficiency ranges from 50% to 90%. This wide variation reflects the diversity of designs and a strong dependence on local climatic conditions.

However, the BOD_5 of effluents can increase in summer due to algae. Between 70 and 90% of the BOD_5 can be measured in an unfiltered sample (Mara and Pescod 1998).

2.2.3.2. *Removal of suspended solids*

In temperate climates, primary lagoons can remove between 60 and 70% of suspended solids from raw water through sedimentation. However, algal growth in facultative lagoons can lead to very high overall concentrations (> 100 mg/l). These high concentrations are mainly recorded in summer in temperate regions (Reed and Zirschky 1988).

The removal of suspended solids is affected by the concentration of settleable solids in the incoming water, the accumulation of sludge and the inversion of the water column.

2.3. Lagoon sizing

An aerated lagoon uses the same principle as a simple lagoon, but with the oxygen supply increased by the installation of mechanical aerators. An aerated lagoon is similar to a large aerobic lagoon operating on the principle of activated sludge at a very low loading rate.

There are two types of aerated lagoons:

– aerobic lagoons in which a concentration of dissolved O_2 is maintained throughout;

– facultative lagoons in which the O_2 is maintained only in the upper part of the lagoon. An anaerobic zone is therefore present at the bottom of the lagoon.

Generally, the sludge is kept in suspension, and this system can be likened to that of activated sludge with a very low loading rate operating without sludge recycling. The treated effluent will therefore have a sludge concentration equal to that present in the lagoon. Secondary settling is essential.

In an anaerobic lagoon, since there are two zones, different phenomena can be expected. During sludge settling, anaerobic degradation leads to products that are then oxidized in the upper layers of the lagoon. This type of process has proven effective in the treatment of some industrial wastewater (slaughterhouses, sugar refineries, for example).

2.3.1. *Applied load*

The applied loading rate can be expressed differently depending on the physical or hydrological factors used. The surface loading approach in BOD_5 (kg $BOD_5.m^{-2}.d^{-1}$ or kg $BOD_5.ha^{-1}.d^{-1}$) is purely empirical, using data from existing lagoons in a particular region. The basis of this approach is that for each climate, there is a specific value that can be applied to a lagoon for a given removal efficiency. In many countries, it is this loading that determines the lagoon size in a particular context. It can represent the BOD_5 loading rate applied per hectare of lagoon, the number of connected inhabitants per hectare of lagoon or the BOD_5 loading removed per hectare or per lagoon volume. It varies from country to country. For anaerobic lagoons, in Australia, for example, it is 500 kg BOD applied/ha.d at 700 kg BOD/ha.d depending on the seasons.

In aerobic oxidation lagoons, the applied loading rate varies from 10 to 40 kg BOD/ha.d, whereas in mixed lagoons, it reaches 20–100 kg BOD/ha.d.

For economic reasons related to energy costs and equipment maintenance, facultative lagoons (without mechanical aeration) are widely used. Since most processes in a facultative lagoon occur at the surface – use of sunlight, algae growth, wind aeration, etc. – it has been found that surface loading values correlate more closely with performance data than volumetric loading values (kg $BOD_5.m^{-3}.d^{-1}$).

Because the volume of the lagoon is not relevant to this approach, the depth of the lagoon can be determined by other factors such as the sludge storage capacity (McGarry and Pescod 1970).

Mara indicates a global design equation for optional lagoons (Mara and Pescod 1988):

$$\lambda s = 350\ (1.107 - 0.002T)^{T-25}$$

where T is the temperature in °C (range 8–25°C).

Other similar relationships have been found, but mainly for warm climates:

$\lambda s = 20\ T - 120$ (Mara and Pescod 1988).

$\lambda s = 20\ T - 60$ (Arthur and Mundial 1983).

$\lambda s = 10\ T$ (Mara and Pescod 1998, for the Middle East).

2.3.2. *Removal of BOD_5*

The lagoons allow for the elimination of 85–90% of the BOD_5 depending on the type of lagoon, the applied loading rate and the effluent retention time. In fact, measuring the BOD_5 often depends on the measurement conditions. For example, for an outlet effluent, analyses revealed a 70–75% removal of the BOD_5 without prior filtration of the sample, and 85–90% after centrifugation. This example demonstrates the importance of operating conditions in the analysis of outlet water.

In fact, the removal of the BOD_5 is no longer a crucial factor here because, while the discharged effluent is devoid of organic matter, it is very rich in living organic matter, represented by algae. While this matter can be quantified, it is not harmful to the receiving environment and does not risk triggering fermentation.

This type of process requires certain essential data to calculate the structure. This includes the number of connected inhabitants, any variations over time, and the composition of the incoming effluent (if industrial effluent is connected to the network).

Boutin and Racault (1974) present a table of the different permissible loading rates depending on climatic conditions, based on a BOD_5 input of 50 g per inhabitant per day (Table 2.1).

Climate	$kgBOD_5$/inhab/d	Population (inhab/ha)	Retention time (d)
Cold zones, ice cover in winter, low water temperature, heavy cloud cover	10	200	200
A cold season characterized by ice cover, temperate climate during a good summer season	10–50	200–1,000	100–200
Temperate to semi-tropical climate, good sunshine	50–150	1,000–3,000	100–33
Uniform tropical climate, no heavy seasonal cloud cover	150–350	3,000–7,000	33–17

Table 2.1. *Allowable loading rates depending on climate conditions*

The water temperature during the coldest month of the year will be recorded, and similarly for the warmest month.

Other parameters can be considered, such as water evaporation during the warmer periods of the year. The desired yield is also important to know when determining the lagoon size.

2.3.2.1. *Calculation of aerated lagoons*

Methods for determining the size of oxidation lagoons have most often been established based on experimental results. The proposed formulae are therefore empirical. Some authors have attempted, using theoretical data, to develop a rational method based on the degradation of organic matter (BOD).

It is accepted that the evolution of the BOD follows an exponential decay law such that:

$$Lf = Lo\ e^{-kt}$$

where k is the BOD_5 degradation coefficient.

The input–output balance of the lagoons gives:

$$Q\ Lo\ dt - Q\ Lf\ dt - V\ Lf\ k\ dt = 0$$

where Lo and Lf are the BOD_5 concentrations at the inlet and outlet of the lagoon, and Q and V represent respectively the inlet flow rate and the lagoon volume.

This equation transforms as follows:

$$Lf = \frac{QLo}{Q + Vk}$$

If:

$$t = V/Q$$

we get:

$$Lf = \frac{Lo}{1 + kt}$$

where k is a temperature-dependent coefficient, such that:

$$k = 1.2 \times 1.085^{(T-10)}$$

2.3.2.2. *Suspended solid equation*

If Q is the inlet flow rate and X is the VSS concentration at the inlet, we can write:

$$Q\,Xo\,dt + \mu\,Xa\,V\,dt - Q\,Xa\,dt = V\,dXa$$

So, by dividing by dt, we obtain:

$$Q\,Xo + \mu\,Xa\,V - Q\,Xa = V\,dXa/dt$$

$$Q\,(Xo - Xa) + \mu\,Xa\,V = V\,dXa/dt$$

where μ is the bacterial growth rate and V is the lagoon volume.

At equilibrium, $dXa/dt = 0$.

Therefore:

$$Q\,(Xo - Xa) = \mu\,Xa\,V$$

Now, beginning with μ:

$$\mu = \frac{Q(Xo - Xa)}{tXo}$$

This shows that the growth rate is inversely proportional to the retention time of the effluent in the lagoon, if the VSS concentration at the inlet (Xo) is significantly lower than that present in the lagoon.

2.3.2.3. *BOD_5 equation*

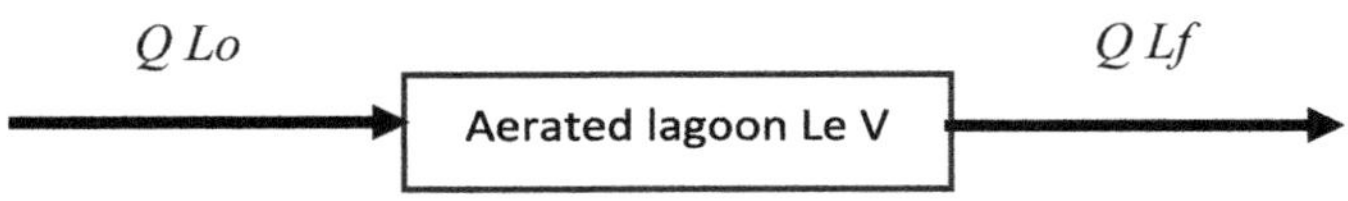

The equation is written as:

$$QLo - QLf + VLe = V\ dL/dt$$

where Le is the removed effluence (BOD $mg.L^{-1}$).

This removed effluence is in principle identical to that of activated sludge, i.e.:

$$am\ Le = \mu Xa$$

which gives:

$$Q\ (Lo - Lf) + \frac{\mu\ Xa\ V}{am} = V\ dL/dt$$

At equilibrium, $dL/dt = 0$.

So, if $t = V/Q$, we get:

$$Lo - Lf = \frac{\mu\ Xa\ t}{am\ Lf}$$

which is written as:

$$Lo/Lf = 1 + \frac{\mu\ Xa\ t}{am\ Lf}$$

And if:

$$k = \frac{\mu\ Xa}{am\ Lf}$$

we finally obtain:

$$\frac{\text{Lo}}{\text{Lf}} = 1 + \text{kt or } \frac{\text{Lf}}{\text{Lo}} = \frac{1}{1 + \text{kt}}$$

The value of k depends on the bacterial growth rate, the biomass concentration in the system, the cell yield and the Lf/Xa ratio.

This relationship only applies in practice if the measured BOD corresponds to a filtered sample representing only dissolved effluence. We saw previously that the outlet effluent could contain a significant concentration of algae, which would skew the treatment results.

The different values of k are presented in Table 2.2.

The mass of sludge produced in the lagoon each day is:

$$\text{V X} = \text{am Le Q} + \text{Q Xo}$$

System	SS	Dissolved oxygen concentration ($mg.L^{-1}$)	k
Simple lagoon (anaerobic)	Low	Zero	0.02–0.10
Aerated lagoon	Medium	Average	0.30–3
Activated sludge	High	Average	2–10

Table 2.2. *k values for different wastewater treatment systems*

Taking into account the fact that there is no recycling, and also endogenous respiration, the equation, over a day, gives:

$$\text{Q Xo} + \text{am Le Q} - \text{b Xa V} = \text{Q Xa}$$

$$\text{Xa} = \frac{\text{amLe+Xo}}{\text{1+bt}} = \frac{\text{amLe}}{\text{1+bt}} \text{ if Xo} \ll \text{Xa}$$

Substituting the value of X into the previous equation, we obtain:

$$\frac{\text{Lf}}{\text{Lo}} = \frac{1}{1 + \dfrac{\text{k}'\text{am Le t}}{1 + \text{bt}}}$$

with $k' = \mu/\text{am Lf}$.

Alternatively, we have:

$$\frac{Lf}{Lo} = \frac{1 + bt}{1 + bt + k'am\ Le\ t}$$

or:

$$\frac{Lo}{Lf} = 1 + \frac{k'am\ Le\ t}{1 + bt}$$

and:

$$\frac{Le}{Lf} = \frac{k'am\ Le\ t}{1 + bt}$$

and so:

$$Lf = \frac{1 + bt}{k'am\ t}$$

This equation implies that the BOD_5 of the treated effluent is not linked to the inlet BOD_5 concentration. It also indicates that there is an increase in the rate of organic matter degradation for increasing inlet BOD values. The active fraction of VSS decreases with sludge age, or in a fully mixed lagoon, with retention time.

This decrease is attributed to the accumulation of inert mass resulting from endogenous phenomena.

The incorporation of a factor x, highlighting this phenomenon, is illustrated by:

$$\frac{Lf}{Lo} = \frac{1}{1 + k\ xXa\ t}$$

x takes a value of 0.75 for young sludge (2 d) and 0.5 for older sludge (6 d). It is easily deduced from this that:

$$xXa = \frac{Le}{Lf\ k't}$$

By replacing k′ with its value, i.e. $k' = (1 + bt)/(am\ Lf\ t)$, we obtain:

$$xXa = \frac{am\ Le}{1 + bt} \text{ and } Lf = \frac{1 + bt}{am\ k't}$$

2.3.2.4. *Oxygen requirements*

The general equation for oxygen requirements is written as:

$$qO_2/t = a'Le\ Q + b'\ Xa\ V$$

which becomes, by replacing Xa with its value:

$$qO_2/t = a'Q\ Le + \frac{b'am\ Le\ V}{1+bt}$$

Or even:

$$qO_2/t = \frac{QLe\ (a'+b'am\ t)}{1+bt}$$

For domestic water, the values of parameters am, a′, b and b′ are:

am = 0.57; a′ = 0.5–0.7; b′ = 1.42 b; b = 0.07–0.10

Many authors believe that endogenous respiration can be disregarded given the low proportion of sludge present in the lagoon. In this case, oxygen requirements become reduced to:

$$qO_2/t = a'\ Le\ Q$$

It is clear that using such an equation involves some risks, given that aerated lagoons often contain a significant amount of sludge which will necessarily consume oxygen.

2.3.2.5. *Sludge production*

Sludge production will depend directly on the difference between the mass of sludge entering the treatment plant and the mass discharged with the effluent, since there is no sludge recycling.

Therefore:

$$\Delta X = Q\ Xa - Q\ Xo$$

Sludge production is also equal to the difference between the amount of living matter synthesized by BOD removal and that due to endogenous respiration:

$$\Delta X = am\ Le\ Q - b\ Xa\ V$$

By equating the two equations, we get:

$$Q\,Xa - Q\,Xo = am\,Le\,Q - b\,Xa\,V$$

We draw from it with t = V/Q:

$$Xa = \frac{Xo + amLe}{1 + bt}$$

If we assume that there is a non-biodegradable fraction of suspended solids in the effluent to be treated (f), and also a volatile fraction of the suspended solids in equilibrium in the lagoon (f'), and that, on the other hand, the concentration of suspended solids in the outlet effluent is not systematically equivalent to that existing in the lagoon, so we can write:

$$X = f\,Xo + \frac{Xa}{t} - Xeff$$

$$Q\,am\,Lo - b\,Xa\,V = Q\,f\,Xo + (V\,Xa/f') - Q\,Xeff$$

i.e.:

$$Xa = \frac{1}{t}\,\frac{(amLe - f\,Xo - Xeff)f'}{1 + bf'}$$

EXAMPLE.– Consider wastewater with an inlet BOD of 550 mg.L^{-1}. We want the output BOD to be 50 mg.L^{-1}. The flow rate is 1,000 m^3 per day.

The incoming water temperature is 30°C, and the air temperature can drop to 2°C. Measuring α gives a value of 0.7, and the quantity of O_2 transferred by turbines is 2 kg O_2/kWh. We want to calculate the oxygen input provided by this type of aerator. Note that the goal is to maintain a dissolved O_2 concentration in the lagoon of 1 mg.L^{-1}.

This gives:

$$a' = 0.65;\ b' = 0.10;\ am = 0.57;\ k = 1.8$$

with an inlet SS of: 60 mg.L^{-1}.

– Calculating retention time:

$$\frac{Lf}{Lo} = \frac{1}{1 + kt} \text{ thus } t = \left(\frac{Lo}{Lf} - 1\right)\frac{1}{k}$$

Alternatively:

$$t = \left(\frac{550}{50} - 1\right)\frac{1}{1.8} = 5.5 \text{ d}$$

The lagoon volume is then: V = 1,000 x 5.5 = 5,500 m^3.

If we adopt a uniform depth of 1.3 m, the surface area is:

S = 5,500/1.3 = 4,230 m^2

The concentration of sludge in the lagoon is:

$$Xa = \frac{X'o - am\ Le}{1 + bt}$$

X′o = 0.65 x 60 = 39 $mg.L^{-1}$ (65% of VSS in inlet SS)

b = b′/1.42 with b = 0.070

$$Xa = \frac{39 + 0.57 \times 500}{1 + 0.07 \times 5.5} = 234 \text{ mg.L}^{-1}$$

Oxygen requirements are:

q O_2/t = a′ Le Q + b′ Xa V

q O_2/t = 0.65 x 0.5 x 1,000 + 0.1 x 0.234 x 5,500

q O_2/t = 453.7 kg $O_2.d^{-1}$

Sludge production is:

Δ X = a′Le Q – b′ Xa V

And so:

Δ X = 0.57 x 0.50 x 1,000 – 0.07 x 0.234 x 5,500 = 195 kg

The temperature of the lagoon is given by:

$$Tl = \frac{QTo + 5.96\ S\ Ta}{Q + 5.96\ S}$$

Flow rate: 1,000 $m^3.d^{-1}$ = 41,666 $l.h^{-1}$

$$\text{Tl} = \frac{41{,}666\text{x}30+5.96\ \text{x}4{,}230\ \text{x}\ 2)}{41{,}666+5.96\ \text{x}4{,}230} = 19.4°\text{C}$$

If we assume that the surface area of the water in contact with the air is multiplied by turbulence and waves, when using mechanical aeration, it is advantageous to multiply the heat loss by a factor f which is approximately 1.5.

The temperature is given by:

$$5.96.\ (\text{Tl} - \text{Ta})\ 1.5\ \text{S} = \text{Q}\ (\text{To} - \text{Tl})$$

i.e.:

$$\text{Tl} = 16.6°\text{C}$$

The aerator's performance is given by:

$$\text{N} = \text{No}\frac{\text{CsT} - \text{C}}{\text{Cs}}\ 1.02^{\text{T}-10}$$

with:

- No = 2 kg O_2/kWh;
- T°: 16.6°C;
- Cs = 10.92 mg.L^{-1} at 10°C;
- CsT = 9.45 mg.L^{-1} = 16.6°C;
- C = 1 mg.L^{-1}.

From this, we can deduce N = 1.79 kg O_2/kWh.

2.3.2.6. *Other models*

An equation relating the concentration of the treated effluent with the lagoon depth has been proposed (Shaw 1962):

$$\text{Lf} = \frac{600}{0.18\text{d} + 8}$$

The lagoon depth H is expressed in meters:

$$\text{Lf} = \frac{\text{Lo}}{0.17\text{t} + 1}$$

Oswald et al. (1967) propose a complex equation involving the heat of combustion (G) of the algae, factor F expressing the efficiency of converted light, solar energy (S) and the concentration of the algae (Cc):

$$t = \frac{G\ Cc\ H}{F\ S\ 1000}$$

The retention time is in days, G is in $cal.g^{-1}$, H is in cm, Cc is in $mg.L^{-1}$ and S is in $g.cal.cm^{-2}.d^{-1}$.

Many other models exist, depending on the type of lagoon, the hydraulic load, the movement of the effluent and the geometry.

2.3.2.6.1. Hermann and Gloyna model

Hermann and Gloyna (1958) developed a kinetic theory to describe the behavior of facultative lagoons. They established that the retention time required for the removal of BOD_5 is a function of temperature. More specifically, the retention time to achieve a 90% reduction in BOD_5 at 35°C was 3.5 days. The retention time required at any other temperature is given by the following relationship:

$$R_T/R_{35} = \Theta^{(35-T)}$$

where:

– Θ = Arrhenius constant = 1.072;

– R_T = retention time required for temperature T;

– R_{35} = 3.5 days.

This model assumes that the BOD in the influent is 200 $mg.L^{-1}$; therefore, with 90% removal, the effluent concentration is 20 mg.L¯1 during the coldest month. For any other BOD value, the calculated retention time must be adjusted in the BOD/200 ratio. The usable range is from 5°C to approximately 35°C, which is the lower limit resulting from the activity of aerobic bacteria and algae.

2.3.2.6.2. Gloyna equation

The Gloyna et al. (1976) equation is an empirical model based on experience from several hundred large-scale lagoons, as well as laboratory and pilot studies. These studies were conducted in the southwestern United States. The equation also includes the influence of temperature (Arrhenius coefficient), as well as two factors related to toxicity for industrial waste or waste high in sulfate:

$$R = 0.035\ La\ \Theta^{(35-T)}\ f.f'$$

where:

– R = hydraulic retention time (d);

– La = BOD ultimate influential ($mg.L^{-1}$);

– Θ = temperature coefficient = 1.085;

– T = water temperature during the coldest month (°C);

– f = algae toxicity factor = 1 for domestic wastewater;

– f' = sulfide or other direct chemical oxygen demand = 1 where sulfate < 500 $mg.L^{-1}$.

For urban wastewater, the equation becomes:

$$R = 0.035\ La\ \Theta^{(35-T)}$$

This can be used to calculate the retention time required for 80–90% removal of BOD_5 (filtered effluent samples). This also shows that the retention time is a function of temperature, BOD concentration in the wastewater influent, algal toxicity and sulfide concentration.

Gloyna suggested that the ultimate BOD be used in the equation because the lagoons must have retention times greater than five days. The ultimate BOD values are generally not available, and therefore sometimes the COD concentration or a multiplier (e.g. 1.5 x BOD_5) are used. The suitable temperature range for the equation is between 5 and 35°C.

The volume of the lagoon is deduced from the relationship:

$$V = RQ$$

where V = required basin volume (m^3) and Q = effluent flow rate ($m^3.d^{-1}$).

Gloyna et al. (1976) proposed a general formula that takes into account the daily volume of water to be treated. Specifically:

$$V = (3.5\ 10^{-5})\ Lo\ Q\ 1.085^{(T-10)}$$

where T = average temperature during the coldest month of the year.

Mechanical oxygenation and mixing can be achieved using surface aerators or diffuse aeration with compressed air. This solution helps maintain sufficient oxygen levels and eliminate odors in lagoons.

Microorganisms are supplied with oxygen by surface mechanical aerators, whereas in normal facultative lagoons, algae provide the oxygen, although aerators can be installed and operated only when needed.

Two types of lagoons using surface aerators are generally used. In aerobic lagoons, all solids are kept in suspension and the concentration of sodium sulfate in the treated effluent is homogeneous.

In facultative aerobic lagoons, the velocities and characteristics of the mixing allow a large proportion of unoxidized biological solids to settle to the bottom. These deposits must undergo anaerobic decomposition (Gloyna et al. 1976).

Surface aerators can be fixed in a stable position or used as floating units. Generally, with fine bubble aeration, 15–20% of the total oxygen required is obtained. Mechanical surface aeration provides approximately 1.5–2.5 kg of oxygen/kW.

The oxygen supplied by mechanical aeration is generally determined as follows:

$$O2s = Om\frac{CsT - C}{Cs}\ \alpha\ F\,\Theta^{T-20}$$

where:

- O_2s = oxygen supplied by the aerators (kg oxygen/kW);
- O_2m = aerator characteristics (supplier data), 1.5–2.5 kg oxygen/kW;
- CsT = saturation concentration in O_2 ($mg.L^{-1}$), at temperature T (°C);
- C = concentration of O_2 in the lagoon ($mg.L^{-1}$);
- Cs = O_2 saturation concentration ($mg.L^{-1}$) in pure water at 20°C;
- A = O_2 transfer coefficient in wastewater;
- Θ = temperature correction coefficient (Θ = 1.02);
- F = correction factor related to an altitude above 1,200 m.

This equation can be used to determine the O_2 requirement in aerobic lagoons with integral mixing.

The O_2 required by microorganisms is calculated as follows:

$$qO_2/t = a'\,(Lo - Lf) + b'\,Xa$$

where:

– qO_2/t = amount of O_2 required ($kg.d^{-1}$);

– a′ = coefficient related to the energy requirements for microorganism growth (0.5–0.7);

– b′ = endogenous respiration coefficient (d^{-1});

– Xa = total SS concentration in the lagoon = Xa V in kg;

– Lo – Lf = BOD_5 removed from the raw wastewater to the outlet effluent = (Lo – Lf) Q in $kg.d^{-1}$.

In lagoons where the amount of SS remains at low concentrations, the second term of the equation cancels out.

Therefore, the equation simplifies as follows:

$$qO_2/t = a' (Lo - Lf)$$

where:

qO_2/t = amount of O_2 required ($kg.d^{-1}$)

EXAMPLE.– Calculation of the lagoon surface area according to the Gloyna method.

– Data:

- Flow rate: 7,570 $m^3.d^{-1}$.

- BOD_5 at the inlet: 250 $mg.L^{-1}$.

- BOD_5 required at the outlet: 30 $mg.L^{-1}$.

- C_{O2} required in the lagoon = 1.5 $mg.L^{-1}$.

- Cs: 9.2 $mg.L^{-1}$.

- C_{ST} = 10.2 $mg.L^{-1}$.

- T = 15°C.

- a′ = 0.7.

- Om = 1.5 kg O_2/kW h at 20°C and 2.14 kgO_2/kW when the dissolved O_2 concentration in the lagoon is negligible.

- α = 0.90.

- kT = 0.35 d^{-1}.

– Solution:

- Applied organic loading rate (kg $BOD_5.d^{-1}$):

  BOD_5 x flow rate = 0.250 x 7,570 = 1,892.5 kg $BOD_5.d^{-1}$

- Retention time:

  RT = (Lo – Lf)/kt Lf = (250 – 30)/0.35 x 30 = 21 d

- Volume:

  Q t = 21 x 7,570 = 158,970 m^3

- Surface area required with a depth of 3 m:

  volume/depth = 158,970/3 = 52,990 m^2

- Oxygen requirement:

  $qO_2/t = a' Le + b' Xa$

  qO_2/t = 0.7 (0.250 – 0.030) 7,570 + 0.07 x 0.5 x 158,970 = 1,165.8 + 5,563.9 = 6,730 kg $O_2.d^{-1}$ = 280.4 kg $O_2. h^{-1}$

- Oxygenation capacity required for surface aerators:

  $qO2/kW = Om\ (Cst - C)/Cs\ \alpha\ \Theta\ (T - 20)$

  $= 1.6 \times [(10.2 - 1.5)/9.2] \times 0.90 \times 1.02^{(15-20)}$

  = 1.23 kgO_2/kW

- Power required:

  280.4/1.23 = 227.6 kW

Another method for calculating lagoon volume (Gloyna and Eckenfelder 1964) is as follows:

$$V = (3.5\ 10^{-5})\ N\ q\ Lo\ \Theta^{(35 - Tm)}$$

where:

– V: lagoon volume (m^3);

– N: connected population (number of inhabitants);

– q: water consumption per capita per day (150 $L.d^{-1}$);

– Θ: temperature correction coefficient = 1.085;

– Tm: average temperature of the coldest month (°C).

The volume and surface area must be increased for wastewater containing more than 500 $mg.L^{-1}$ of sulfate ions (Gloyna and Espino 1969). Similarly, a larger volume is required for certain industrial wastewaters because the organic substances present are not easily biodegradable, because they are toxic to algae, or because they reduce chlorophyll content (Huang and Gloyna 1968).

EXAMPLE.–

– Data:

- BOD_5 at the inlet: 250 $mg.L^{-1}$.
- BOD_5 required at the outlet: 30 $mg.L^{-1}$.
- Connected population (N): 50,466.
- Consumption per capita: 150 liters per day.
- Flow rate (Nq): 7,570 $m^3.d^{-1}$.
- Tm: 10°C.
- Θ: 1.085.
- Sulfate concentration: < 500 $mg.L^{-1}$.
- Concentrations of toxic industrial products: negligible.

– Solution:

- Applied organic load:

  $Q\ (m^3.d^{-1}) \times BOD_5 \text{ inlet } (kg.m^{-3}) = 0.250 \times 50{,}466 \times 0.150 = 1{,}892.4 \text{ kg } BOD_5.d^{-1}$

- Volume:

  $$V = (3.5\ 10^{-5})\ N\ q\ Lo\ \Theta^{(35 - Tm)}$$

  $$= (3.5\ 10^{-5}) \times 50{,}466 \times 150 \times 250 \times 1.085^{(35 - 10)}$$

  $$= 509{,}145\ m^3$$

- Depth: 3 m.
- Required surface area:

  volume/depth = 509,145/3 = 169,175 m^2 = 16.9 ha

- Retention time:

volume/flow rate ($m^3.d^{-1}$) = 509,145/7,570 = 67 d

- Volumetric loading rate applied per lagoon volume:

7,570 x 0.250/509,145 = 0.0037 kg BOD.$m^{-3}.d^{-1}$

- Loading rate applied per lagoon surface:

7 570 x 0.250/169,175 = 0.0112 kg $BOD_5.m^{-2}.d^{-1}$ = 112 kg $BOD_5.ha^{-1}.d^{-1}$

The difference stems from the factor 3.5 10^{-5} which overestimates the volume calculation. We find the previous volume value again by taking $V = (1.1\ 10^{-5})\ N\ q\ Lo\ \Theta^{(35-Tm)}$.

2.3.2.6.3. Marais method

Marais and Shaw (1961) proposed a model based on first-order kinetics in an integrally mixed lagoon. In this configuration, the BOD_5 concentration is constant and equal to the BOD_5 concentration of the treated effluent:

$$\frac{Cn}{Co} = \left(\frac{1}{1 + kc\ rn}\right)^n$$

where:

- Cn: BOD_5 of the treated effluent from the lagoon ($mg.L^{-1}$);
- Co: BOD_5 influent ($mg.L^{-1}$);
- kc: first-order reaction rate constant (d^{-1});
- rn: hydraulic retention time in each lagoon (d);
- n: number of lagoons in a series.

The value of kc depends on the temperature as follows:

$$kc_T = kc_{35}\ 1.085^{(35-T)}$$

where:

- kc_T: reaction rate at the minimum service water temperature;
- kc_{35}: reaction rate at 35°C = 1.2 d^{-1};
- T: minimum operating water temperature (°C).

The equation can be used to determine both the retention time and the number of lagoons. Marais proposes that the maximal BOD_5 in primary cells ((Ce)max), to avoid anaerobic conditions and odors, must be 55 $mg.L^{-1}$. The required retention time in the first lagoon can be calculated by replacing (Ce)max with Cn in the previous equation where n = 1.

The permitted lagoon depth is given by:

$$(\text{Ce})\text{max} = \frac{700}{2\text{d} + 8}$$

where d is the lagoon depth (m).

Using this equation, when (Ce)max = 55 $mg.L^{-1}$, d = 2.4 m.

Mara points out that the equation is more sensitive to its numerator (600, the choice of which is arbitrary) than to the depth (Mara 1975).

Another formulation separates the BOD_5 calculation according to the operating zone. Lf_1 is the BOD_5 concentration at the outlet of the first lagoon ($mg.L^{-1}$).

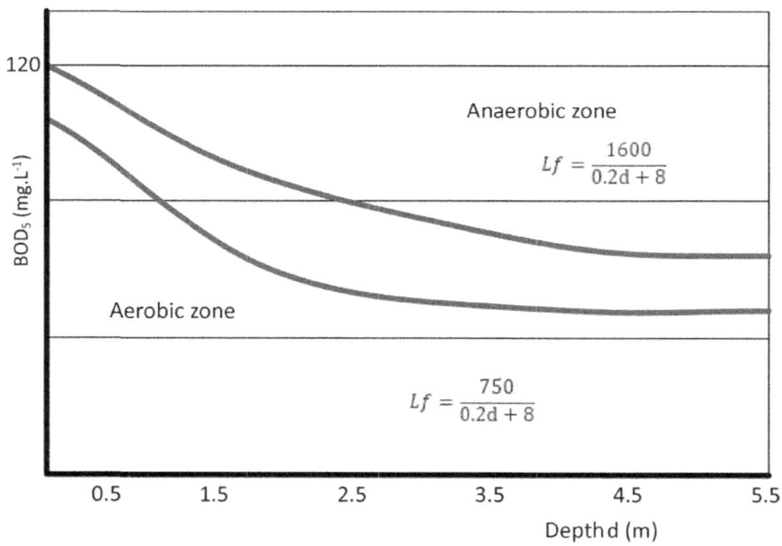

Figure 2.9. *BOD_5 concentration in the aerobic lagoon for different lagoon depths*

BOD_5 represents the expected value in the treated effluent (see Figure 2.9).

In South Africa, the maximum load in kg $BOD_5.d^{-1}$ for an optional lagoon is linked to the depth by the relationship:

$$Lf = \frac{600}{0.18\ d + 8}$$

An equation that takes into account the retention time gives:

$$Lf = \frac{Lo}{0.18\ RT + 1}$$

The quality of effluents treated by lagoons requires a BOD_5 value of 60 $mg.L^{-1}$, for a depth of 1.75–2 m. The equations proposed by Marais are applicable under these conditions to determine lagoon surface area. When the value reached is not acceptable, additional secondary or even tertiary treatment is implemented.

These additional lagoons are designed for a single retention time of 7 days or more. In this case, the BOD_5 of the effluents from the lagoons arranged in series does not exceed 20 $mg.L^{-1}$. The surface area of each lagoon is calculated individually based on the applied pollutant loads.

EXAMPLE.– Facultative lagoons.

– Data:

- BOD_5 at the inlet: 250 $mg.L^{-1}$.
- BOD_5 required at the outlet: 60–072 $mg.L^{-1}$.
- Linked population (N): 50,466 inhabitants.
- Consumption per capita: 150 liters per day.
- Flow rate (Nq): 7,570 $m^3.d^{-1}$.
- Tm: 10°C.
- Θ: 1.085.
- KT: 0.17 d^{-1}.
- Lagoon depth: 1.75 m.

– Solution:

- Estimated BOD_5 at the outlet ($mg.L^{-1}$):

$$Lf = \frac{600}{0.18\ d + 8} \text{ i.e. } Lf = \frac{600}{0.18 \times 1.75 + 8} = 72$$

- Retention time (d):

$$\mathrm{RT} = (\frac{\mathrm{Lo}}{\mathrm{Lf}} - 1)(\frac{1}{\mathrm{KT}}) \text{ i.e. } \mathrm{RT} = \left(\frac{250}{72} - 1\right)\left(\frac{1}{0.17}\right) = 14.5 \text{ d}$$

- Volume required (m^3):

Q t = 7,570 x 14.5 = 109,765

- Surface required:

volume/depth = 109,765/1.75 = 62,722 m^2 = 6.27 ha

- Volumetric loading rate applied per lagoon:

7,570 x 0.250/109,765 = 0.017 kg BOD.$m^{-3}.d^{-1}$

- Loading rate applied per lagoon surface area:

7,570 x 0.250/62,722 = 0.030 kg $BOD_5.m^{-2}.d^{-1}$ = 301 kg $BOD_5.ha^{-1}.d^{-1}$

2.3.2.6.4. Methodology of the CPHERI, India

In tropical India, the design of facultative waste stabilization lagoons for domestic wastewater is based on a load of 336 kg BOD_5/ha per day and an operating depth of 1.22 m (Dave and Jain 1967). A study of eight lagoons showed a 70–85% reduction in the BOD_5 for this loading rate. This organic load is considered inaccessible in many geographical regions, even though summer temperature conditions are similar to those observed in India. To reach 90% removal of BOD_5, the retention time of wastewater in the lagoons must be significantly increased (Gloyna 1971).

EXAMPLE.–

– Data:

- Population: 50,266 inhabitants.
- Water consumption per capita: 227 liters per capita per day.
- Flow rate: 11,410 $m^3.d^{-1}$.
- BOD_5 at the inlet: 250 $mg.L^{-1}$.
- Suspended solids (SS): 300 $mg.L^{-1}$.
- Length/width ratio: 2.
- Depth: 1.22 m.

– Solution:

- Applied organic loading:

  $11{,}410 \times 0.250 = 2{,}852\ 6$ kg $BOD_5.d^{-1}$

- Organic loading per hectare: 336 kg BOD5/ha/d;

- Surface area of the lagoons:

  $2{,}852.6/336 = 8.48$ ha $= 84{,}898.8\ m^2$

- Volume:

  $84{,}898.8\ m^2 \times 1.22 = 103{,}576\ m^3$

- Retention time:

  $103{,}576/11{,}410 = 9.1$ d

It is difficult to compare all values based on the same assumptions (population, flow rate, BOD_5) because the relationships are empirical and depend on operating conditions (applied load, depth, temperature) and hydraulic conditions (piston flow, integral mixing, dispersed flow).

2.3.2.6.5. Piston flow model

A piston flow system is a theoretical concept in which each element of the flow leaves the lagoon in the same direction it entered. Thus, each element is exposed to treatment for the same hydraulic retention time. The lagoon performance model in series is given by (Reed and Zirschky 1988):

$$\frac{Ce}{Co} = EXP\ (-kp\ R)$$

where:

– Ce: BOD_5 concentration in the effluent ($mg.L^{-1}$);

– Co: BOD_5 concentration in the affluent ($mg.L^{-1}$);

– kp: first-order reaction rate at the piston flow (d^{-1});

– R: hydraulic retention time (d).

kp is related to temperature as follows:

$kp = kpT\ 1.09^{(T-20)}$

where:

– kpT: reaction rate at temperature T;

– kp_{20}: reaction rate at 20°C;

– T: minimum wastewater temperature (°C).

kp_{20} depends on the loading rate BOD_5 as shown in Table 2.3, but, if this case is not listed, a value of 0.1/d is suggested.

Kg/ha/d	kp20/d
22	0.045
45	0.071
67	0.083
90	0.096
112	0.129

Table 2.3. *Correspondence between kp values and applied loadings*

The plug flow model can be used to calculate the retention time required for BOD removal. The required volume is derived from the flow rate and retention time. The basin surface area is calculated for a given depth.

2.3.2.6.6. Dispersed flow

The Wehner–Wilhelm equation was proposed by Thirumurthi (1974) as a method for sizing facultative lagoons:

$$\frac{Ce}{Co} = \frac{4\,a\,e^{1/2D}}{(1+a)^2 e^{a/2D} - (1-a)^2 e^{-a/2D}}$$

where:

– Ce: BOD_5 concentration of effluent ($mg.L^{-1}$);

– Co: BOD_5 concentration of influent ($mg.L^{-1}$);

– $a = \sqrt{(1 + kRD)}$;

– k: first-order reaction constant (d^{-1});

– R: hydraulic retention time (d);

– D: dispersion number = H/vl = Ht/l^2;

– H: axial dispersion coefficient (area/time);

– V: circulation rate (length/time);

– L: journey length;

– kT: reaction rate constant at the lowest temperature;

– k_{20}: reaction rate constant at 20°C = 0.5/d;

– T: minimum wastewater temperature (°C).

We obtain:

$$kT = k_{20}\ 1.09^{(T-20)}$$

2.3.3. *Retention time*

In aerobic or facultative lagoons, the effluent retention time varies from 30 to 60 days depending on whether the incoming effluent is settled or discharged directly into the basin (without settling). The retention time in anaerobic lagoons varies seasonally between 2.5 and 3 days in summer, and 5 and 7 days in winter.

These variations depend on various physical factors such as light and temperature. The advantage of a long retention time is that it makes it possible to withstand sudden load variations, which is beneficial for cities with large population fluctuations (coastal cities, for example).

2.3.4. *Ammonia removal from facultative lagoons*

It is assumed that the removal of NH_3 in facultative wastewater treatment basins occurs via three mechanisms: capture of gaseous NH_3 in the atmosphere, assimilation of NH_3 into algal biomass and biological nitrification. Low concentrations of NO_3^- and NO_2^- measured in pond effluents indicate that nitrification generally does not represent a significant portion of the NH_3 removed. The assimilation of ammonia in algal biomass depends on the biological activity of the system and is affected by temperature, organic load, retention time and wastewater characteristics. The release rate of NH_3 gas into the atmosphere depends primarily on the pH value, temperature and mixing conditions in the lagoon. Alkaline pH shifts the equilibrium equation, while mixing conditions affect the mass transfer coefficient:

$$NH_3\uparrow + H_2O \leftrightarrow NH_4^+ + OH^- \text{ towards } NH_3 \text{ gas}$$

Temperature influences both the equilibrium constant and the mass transfer coefficient.

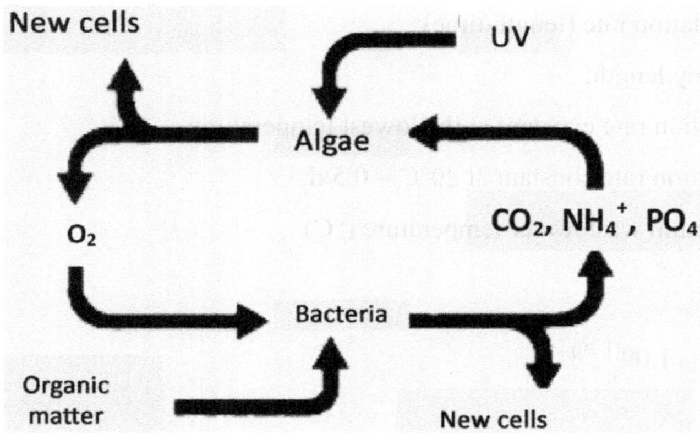

Figure 2.10. *Bacteria–algae biochemical cycle*

At low temperatures, when biological activity decreases and the contents of the lagoon are generally well mixed due to the effects of the wind, stripping will be the dominant process in the removal of NH_3 in wastewater treatment lagoons. The NH_3 removal process from lagoons can be expressed by assuming a first-order reaction (Stratton 1969).

The mass balance equation will be:

$$VdC/dt = Q\ (Co - Ce) - kA\ (NH_3)$$

with:

– Q: flow rate (m^3/d);

– Co: $N–NH_4^+$ concentration at the inlet ($mg.L^{-1}$);

– Ce: $N–NH_4^+$ concentration at the outlet ($mg.L^{-1}$);

– C: average concentration of the lagoon contents in the form of $N–NH_4^+$ ($mg.L^{-1}$);

– V: lagoon volume (m^3);

– k: mass transfer coefficient ($m.d^{-1}$);

– A: lagoon surface area (m^2);

– t: time (d).

At equilibrium, the dissociation reaction of NH_3 can be written as follows:

$$Kb = [NH_4^+]\,[OH^-]/[NH_3]$$

where Kb = NH_3 dissociation constant.

The concentration of NH_3 gas can be expressed as a function of pH and total concentration (NH_4^+ and NH_3):

$$[H^+] = Kw/[OH^-]$$

$$C = NH_4^+ + NH_3$$

$$NH_3 = \frac{C}{10^{pKw-pKkb-pH}}$$

with:

$$pKw = -\log Kw$$

$$pKb = -\log K$$

At equilibrium, Ce = Co.

We can deduce the equation:

$$\frac{Ce}{Co} = \frac{1}{1 + \frac{AK(1 + 10^{pKw-pKkb-pH})}{Q}}$$

NH_3 removal is described as a first-order kinetic process in a lagoon operating under piston flow:

$$Nt = No\ e^{-0.03t}$$

where Nt is the total N concentration (mg.L^{-1}) at time t and No is the initial total in N concentration (mg.L^{-1}). The time t is in d.

The equation for the removal of Kjeldahl nitrogen (TKN) in facultative lagoons is identical to the theoretical expression for NH_3 stripping:

$$\frac{Ce}{Co} = \frac{1}{1 + \frac{AK(f(pH))}{Q}}$$

with:

– K: coefficient linked to the removal rate (L/t);

– f (pH): pH function;

– A: lagoon surface area (m^2).

K values are considered to be a function of temperature and mixing conditions. The pH function, considered to be temperature-dependent, affects the values of pK and pKb, as well as biological activity. Based on a statistical analysis of the data incorporated into the equation, the pH function describes an exponential relationship. Since many relationships between reaction rate and temperature are described by exponential functions, such as the Van't Hoff–Arrhenius equation, it is reasonable to assume that such a relationship applies in the same way.

2.3.4.1. *Reed model*

The Reed model (1985) is as follows:

$$Ne = No\ Exp\ (-K_T(t + 60.6(pH - 6.6))$$

where:

– Ne: total nitrogen in the treated effluent ($mg.L^{-1}$);

– No: total nitrogen at the inlet ($mg.L^{-1}$);

– K_T: temperature-dependent rate constant with $K_T = K_{20}\ \theta^{(T-20)}$;

– K_{20}: rate constant at 20°C = 0.0064;

– θ: 1.039;

– t: retention time (d);

– pH: pH in the lagoon = $7.3e^{0,0005 Alkalinity}$.

2.3.4.2. *Middlebrooks complete mix model*

The Middlebrooks (1982) model is as follows:

$$Ne = \frac{No}{1 + t(0.000576T - 0.00028)e^{(1.080 - 0.0042T)(pH - 6.6)}}$$

Pano and Middlebrooks (1982) proposed a model for ammonia removal incorporating hydraulic head, pH, temperature and coefficients obtained from empirical data:

$$C = C \{1 + [(A/Q)\ (0.0038 + 0.000134\ T)\ \mathrm{Exp}\ ((1.041 + 0.044T)\ (pH - 6.6))]\}$$

where:

– Ce and Co: ammonia concentrations respectively in the lagoon effluent and influent (mg $N.L^{-1}$);

– Q: wastewater flow rate ($m^3.d^{-1}$);

– A: lagoon surface area (m^2);

– T: temperature (°C).

The equation assumes first-order removal kinetics, complete mixing and high temperatures of 20°C. A good fit is achieved with lagoon systems in temperate regions, although the model was developed using three systems with extreme wastewater characteristics.

2.3.5. *Light*

Light is an essential factor in photosynthesis; algae are microscopic aquatic plants that use light energy to transform carbon dioxide, ammonia and mineral elements or matter into new oxygen-releasing cells. It is clear that at night, algae consume the oxygen necessary for their survival. Light therefore influences algae production, which in turn affects the lagoon's efficiency. Generally, water depth determines algae growth, as shallow water promotes light penetration. If the water depth is too high, anaerobic layers will form.

2.3.6. *Microbial decontamination*

The presence of bacteria can be considered a pollution indicator, especially when the health impact of the discharge into the receiving environment is significant.

The processes of bacterial elimination are expressed in the following model:

$$\frac{Nt}{No} = \frac{1}{1 + kt}$$

where No and Nt are the numbers of test bacteria after a retention time t, and k is the decontamination rate (d^{-1}), also called the germ reduction rate.

Fecal contamination test germs often see a reduction close to 100% (Enterococci and *Escherichia coli*).

Several hypotheses have been put forward to explain this phenomenon, such as a possible release of toxic substances by algae, extreme competition between organisms, the role of oxygen saturation, the germicidal action of light, etc.

The advantage of such a technique is that it relieves water treatment companies of the risk of potential microbial contamination of groundwater used in public water supplies.

2.4. Construction of oxidation lagoons

The construction of lagoons involves various factors including location, the influence of several operating parameters and the operation of the treatment plant.

2.4.1. *Implementation*

Field studies (topography and soil type) are necessary. It is clear that installing such a lagoon on permeable ground would not be beneficial to the process.

The direction of the prevailing winds must also be taken into account.

2.4.2. *Sizing and equipment*

The lagoons are generally installed on pre-compacted soil. Dikes are built using the excavated material, and they must be at least two meters wide to allow passage of construction equipment. Growing grass around the edges of the pond is recommended to prevent erosion from waves during strong winds. However, this practice may encourage the development of mosquitoes.

After screening, the wastewater arrives via a supply channel which includes a flow measurement device that is essential during control measurements.

The water supply is generally provided by a pipe located at the bottom of the lagoon, at a distance that varies according to different sources: between one-third and one-half of the lagoon's length.

In anaerobic lagoons, water is often admitted above the water surface to avoid clogging the pipe with accumulated sludge.

Sometimes, the incoming flow is divided into two horizontal pipes. In secondary lagoons, a settling tank follows the screening structure. The depth of the pond varies depending on whether the lagoon is aerobic or anaerobic.

Sludge removal depends on the type of lagoon. Anaerobic lagoons are emptied much more frequently than aerobic lagoons.

The emptying frequency depends on the composition of the incoming effluent (sand, stones, etc.). For example, in Melbourne, the anaerobic lagoon is emptied every 1.5–2 years, while the aerobic lagoon is emptied every 10 years. There are aerobic basins that have been operating for 20 or 30 years and have never been emptied.

The two major drawbacks of oxidation lagoons are the proliferation of mosquitoes and the release of bad odors.

Anaerobic lagoons, particularly those treating water laden with sulfates, are the primary source of odors.

When they are well-lit and oxygenated, aerobic lagoons give off little odor. Cloudy skies for several consecutive days weaken photosynthetic processes, creating anaerobic zones whose fermentation products lead to unpleasant smells.

As a precaution, lagoons should be located a good distance from houses. The proliferation of aquatic plants encourages mosquito breeding. Regularly cutting back these plants should help reduce the mosquito population. Often, insecticide spraying is also carried out.

In recent years, particularly in the United States, attempts have been made to recover the removed algae, which, after selection of certain species, are used as livestock feed.

Furthermore, oxidation lagoons can be used as fish farming ponds, and results obtained in Munich seem promising, with nearly 300 kg of carp produced per hectare of lagoons.

2.4.3. *Influence of temperature*

As in all biological processes, temperature will influence the growth rate of bacteria; in lagoons, temperature variations will affect the system's performance. The lagoon temperature depends on the temperature of the water being treated and the surrounding air temperature. Therefore, heat losses can be significant, and they are due to multiple causes: evaporation, convection and radiation.

The total heat lost in such a system is:

$$H = He + Hc + Hr - Hs$$

where He is the heat loss due to evaporation, He that due to convection, Hr that due to radiation and Hs the heat supplied by light rays from the sun.

He can be given by:

$$He = 0.0195\ Hv\ C\ (1 - 0.1W)\ (V_l - Va)$$

where Hv is the heat of evaporation, C is a constant characteristic of the lagoon, W is the wind speed, V_l is the vapor pressure over the lagoon and Va is the vapor pressure in the ambient air.

Hc is given by:

$$Hc = (2.2 + 1.05W/2)\ (T_l - Ta)$$

where T_l and Ta are the temperatures in the lagoon and the air (in °C) respectively.

Hr is obtained from the following:

$$Hr = 2.7\ (T_l - Ta)$$

In fact, the most significant heat losses are those due to convection and radiation:

$$H = Hc + Hr$$

$$H = 3.25\ (T_l - Ta) + -2.71\ (T_l - Ta)$$

Hence:

$$H = 5.96\ (T_l - Ta) \text{ expressed in cal.m}^{-2}\text{.h}^{-1}$$

The heat balance is therefore, for an inlet flow rate Q ($L.h^{-1}$):

$$5.96.\ (T_l - Ta) = Q\ (To - T_l) \text{ and } T_l = (Q\ To + 5.96\ S\ Ta)/(Q + 5.96\ S)$$

where S is the surface area of the lagoon, and To and T_l represent the inlet and outlet effluent temperatures (lagoon) respectively.

2.4.4. *Lagoon aeration*

During aeration, two essential factors come into play:

– the relative saturation of the effluent;

– the oxygen transfer coefficient.

Despite the significant natural aeration of the water, which depends on various factors (lagoon geometry, surface area, etc.), artificial oxygenation is provided using surface aerators.

The specific contribution of aerators remains significantly lower than that obtained in activated sludge systems due to the flattened shape of the lagoon, and also to the low specific power, which rarely exceeds 3 W/m^3.

The required power must take into account the spacing of the aerators, the depth of the lagoon and the power of each aerator.

It is worth remembering that aeration also allows for mixing of the entire system, preventing sedimentation and the transition to anaerobic conditions at the bottom of the lagoon. While lagoon treatment remains an extensive process, it is nonetheless a controlled method offering practical advantages for wastewater treatment, particularly for industrial wastewater.

2.4.5. *Secondary settling*

The size of settling tanks is determined based on the various constraints required for wastewater treatment. Thus, the settling tank must allow sufficient retention time for the effluent to ensure the removal of suspended solids. It must be neither undersized, which would result in a loss of efficiency, nor oversized, as this would lead to an acceleration of the anaerobic reactions of the settled sludge, resulting in sludge rising to the surface and odors.

2.4.6. *Exploitation*

The construction process is essentially identical to that of oxidation lagoons. The major problem remains the potential contamination of groundwater. The soil type must be carefully studied before choosing the site, although an impermeable lining is often installed. In this case, the lagoon loses all its economic value, especially since the lining can be made of various materials such as concrete, plastic films, bituminous sheets, etc. Erosion of the bottom must also be avoided, particularly when the soil lacks the desired mechanical strength.

Odor problems are practically nonexistent in aerated lagoons, especially when the water level is at least one meter above the bottom. In summer, when temperatures are high, it is advisable to maintain a water level of two meters.

Algal growth can also disrupt the process, as it results in an increase in the output BOD, at least during certain times of the year.

2.5. Conclusion

Advantages	A very simple and easy-to-operate process No electricity is required once the effluent enters the treatment plant; resulting in energy savings Good nutrient removal efficiencies (total nitrogen and total phosphorus) This process is suitable for combined sewer systems that can handle diluted effluent and hydraulic overloads Effective removal of pathogens in summer (provided there is little algae and duckweed growth on the surface)
Disadvantages	Very large footprint Average removal of organic matter (COD, BOD_5) Variable effluent quality depending on the season High sensitivity to concentrated and/or septic effluents (bacterial turnover, foul odors) Potentially significant discharge of solids, especially with the development of algae and duckweed (exacerbated by climate change) Significant occasional operational constraints (liquid sludge removal), operational constraints after removal in case of membrane damage Limited control of the biological balance and treatment processes (Besnault and Prost-Boucle 2024)

Table 2.4. *Summary of advantages and disadvantages of the lagoon system*

In the field of wastewater treatment, lagoon systems offer a robust, economical and environmentally friendly solution. These systems, also known as pond systems, are a natural wastewater treatment method that has been used for centuries. They are particularly popular in rural areas and small communities due to their simplicity and low maintenance requirements.

Despite the apparent simplicity, lagoons are complex ecosystems that rely on a variety of biological, chemical and physical processes to purify wastewater. Understanding these processes and the factors that influence them is essential for the effective operation and maintenance of a lagoon system.

2.6. References

Abbas, H., Nasr, R., Seif, H. (2006). Study of waste stabilization pond geometry for the wastewater treatment efficiency. *Ecological Engineering*, 28, 25–34.

Agunwamba, J.C. (1992). A new method for dispersion number determination in waste stabilization pond. *Water Air and Soil Pollution*, 63, 361–369.

Alexandre, O., Boutin, C., Duchène, P., Lagrange, C., Lakel, A., Liénard, A., Orditz, D. (1997). Filières d'épuration adaptées aux petites collectivités. Technical Document, FNDAE.

Alvarado, A., Sanchez, E., Durazno, G., Vesvikar, M., Nopens, I. (2012). CFD analysis of sludge accumulation and hydraulic performance of a waste stabilization pond. *Water Science and Technology*, 66, 2370–2377.

Arceivala, S.J. (1983). Hydraulic modeling for waste stabilization ponds. *Journal of Environmental Engineering – ASCE*, 109, 265–268.

Arthur, J.P. and Mundial, B. (1983). Notes on the design and operation of waste stabilization ponds in warm climates of developing countries. Technical Report, World Bank, Washington, D.C.

Bastos, R.K.X., Cabral, V.A.L., Rios, E.N., Combatt, M.P.M. (2014). Further contributions to nitrogen removal modelling in waste stabilization ponds. *Water Science and Technology*, 70, 1897–1906.

Beran, B. and Kargi, F. (2005). A dynamic mathematical model for wastewater stabilization ponds. *Ecological Modelling*, 181, 39–57.

Besnault, S. and Prost-Boucle, S. (2024). Réhabilitation des lagunes. Report, INRAE – UR REVERSAAL.

Boettcher, P. and Viall, J. (2020). Basic Lagoon Wastewater Treatment Manual. Montana Department of Environment Quality.

Boutin, P. (1976). *Le lagunage*. Ecole nationale de la santé publique, Rennes.

Boutin, P. (1981). *Le lagunage, point de repère*. A.G.H.T.M., Paris.

Boutin, P. and Racault, Y. (1974). Le lagunage naturel, situation actuelle d'une technique d'épuration en France. *Techniques Sciences Méthodes*, 81(6), 273–284.

Boutin, P., Bechac, J.P., Mercier, B., Nuer, P. (1983). *Traitement des eaux usées*. Ed. Eyrolles, Paris.

Canter, L.W. and Englande, A.J. (1970). States design criteria for waste stabilization ponds. *Journal Water Pollution Control Federation*, 42, 1840–1847.

Dave, J.M. and Jain, J.S. (1967). Simplified design, construction and maintenance of oxidation ponds for practicing engineers. *Instn. Engrs. India*, 47(10), 132–142.

Durot, M.A. and Molle, P. (2015). Amélioration du rejet des lagunes d'épuration. Research Report, Irstea.

Eckenfelder, Jr., W.W. (1982), Industrial wastewater treatment technology – Recent developments. *Water Science and Technology*, 14(1–2), 111–119.

El Ouarghi, H. (2002). Interaction phytoplancton qualité physico-chimique des eaux dans les lagunes naturelles et lagunes à haut rendement algal. PhD Thesis, Fondation Universitaire Luxembourgeoise (FUL), Arlon.

Environmental Protection Agency (EPA) (1983). Municipal wastewater stabilization ponds. Design Manual, United States Environmental Protection Agency, Office of Research and Development, Cincinnati, OH.

Environmental Protection Agency (EPA) (1987). Report on the use of wetlands for municipal wastewater treatment and disposal. Report, EPA 430/09–88-005, Office of Municipal Pollution Control.

Environmental Protection Agency (EPA) (2000). Constructed wetlands treatment of municipal wastewaters. Report, EPA/625/R-99/010, Office of Research and Development, Cincinnati, OH.

Environmental Protection Agency (EPA) (2011). Principles of design and operations of wastewater treatment pond systems for plant operators, engineers, and managers. Report, EPA/600/R11/088, Office of Research and Development, Cincinnati, OH.

Ferrara, R.A. and Harleman, D.R.F. (1981). Hydraulic modeling for waste stabilization ponds. *Journal of the Environmental Engineering Division*, 107, 817–830.

Finney, B.A. and Middlebrooks, E.J. (1980). Facultative waste stabilization pond design. *Journal Water Pollution Control Federation*, 52, 134–147.

Frank, R.S. and Joanne, E.D. (2014). *Wastewater Stabilization Ponds*. CRC Press, Boca Raton, FL.

Gloyna, E.F. (1971). *Waste Stabilization Ponds*. World Health Organization, Geneva.

Gloyna, E.F. and Eckenfelder, W.W. Jr. (eds) (1964). Advances in water quality improvement. In *Water Resources Symposium No. 1*. University of Texas Press, Austin.

Gloyna, E.F. and Espino de la O,E. (1969). Sulfide production in waste stabilization ponds. *Journal of the Sanitary Engineering Division ASCE*, 98(SA3), 607.

Gloyna, E.F., Malina, J.F., Davis, E.M. (1976). *Ponds as a Wastewater Treatment Alternative*. Center for Research in Water Resources, College of Engineering, University of Texas at Austin, Austin, TX.

Gujer, W. (2006). Activated sludge modelling: Past, present and future. *Water Science and Technology*, 53, 111–119.

Heaven, S., Salter, A.M., Clarke, D. (2011). Calibration of a simple model for waste stabilisation pond performance in seasonal climates. *Water Science and Technology*, 64, 1488–1496.

Hermann, E.R. and Gloyna, E.F. (1958). Waste stabilization ponds, experimental investigations, formulation of design equations. *Sewage and Industrial Wastes*, 39, 511–645.

Herrera, L. and Castillo, G. (2000). Dynamics of a waste stabilization pond in Chile. *Water Science and Technology*, 42, 83–90.

Ho, L., Van Echelpoel, W., Goethals, P. (2017). Design of waste stabilization pond systems: A review. *Water Research*, 15(123), 236–248.

Huang, J. and Gloyna, E.F. (1968). Effect of organic compounds on photosynthetic oxygénation. Part II. Design modification for waste stabilization ponds. *Water Research*, 2, 459–471.

Liénard, A., Boutin, C., Molle, P., Racault, Y., Brissaud, F., Picot, B. (2004). Filtres plantés de roseaux à flux vertical et lagunage naturel en traitement d'eaux usées domestiques en France, FNDAE. *Ingénierie*, 87–99.

Lukman, S., Otun, J., Ismail, A., Adie, D., Oke, I., Abubakar, U. (2009). Application of traditional & modern approaches to waste stabilization pond modelling & design. In *6th National Conference of Society for Occupational Safety and Environmental Health*. Nnamdi Azikiwe University, Awka, 13–25.

Mara, D.D. (1975). Proposed design for oxidation ponds in hot climates. *Journal of the Environmental Engineering Division – ASCE*, 101, 296–300.

Mara, D.D. (2009). Waste stabilization ponds: Past, present and future. *Desalination and Water Treatment*, 4, 85–88.

Mara, D.D. and Pescod, M.B. (1988). Design, operation and maintenance of waste stabilization ponds. In *Waste Stabilization Ponds: Design Manual for Mediterranean Europe*, Mara, D.D. and Pearson, H.W. (eds). World Health Organization Regional Office for Europe, Copenhagen, 94–118.

Mara, D.D. and Pearson, H.W. (1998). *Waste Stabilization Ponds: Design Manual for Mediterranean Europe*. World Health Organization, Regional Office for Europe, Copenhagen.

Marais, G.V.R. (1970). Dynamic behaviour of oxidation ponds. In *Proceedings of the Second International Symposium on Waste Treatment Lagoons*, Kansas City, 15–46.

Marais, G.V.R. and Shaw, V. (1961). A rational theory for the design of sewage stabilization ponds in central and South Africa. *Transactions*, 3, 205–218.

Mayo, A.W. (1995). Modeling coliform mortality in waste stabilization ponds. *Journal of Environmental Engineering – ASCE*, 121, 140–152.

Mayo, A.W. and Abbas, M. (2014). Removal mechanisms of nitrogen in waste stabilization ponds. *Physics and Chemistry of the Earth*, 72–75, 77–82.

McGarry, M.G. and Pescod, M.B. (1970). Stabilization pond design criteria for tropical Asia. In *Proceedings of the Second International Symposium on Waste Treatment Lagoons*, Kansas City, 114–132.

McGarry, M.G., Lin, C.D., Merto, J.L. (1973). Photosynthetic yields and byproduct recovery from sewage oxidation ponds. *Advances in Water Pollution Research*, 4, 521–531.

Metcalf & Eddy (1991). *Wastewater Engineering: Treatment, Disposal and Reuse*, 3rd edition. McGraw-Hill, Inc., Singapore.

Middlebrooks, E.J. and Pano, A. (1983). Nitrogen removal in aerated lagoons. *Water Research*, 17(10), 1369–1378.

Middlebrooks, E.J., Middlebrooks, C.H., Reynolds, J.H., Walters, G.Z., Reed, S.C., George, D.B. (1982). *Wastewater Stabilization Lagoon Design, Performance and Upgrading.* Macmillan, New York.

Middlebrooks, E.J., Reed, S.C., Pano, A., Adams, V.D. (1999). Linking sanitation and water treatment. In *6th National Drinking Water and Wastewater Treatment Technology Transfer Workshop*, Kansas City, 2–4 August.

Oswald, W.J., Golueke, C.G., Tyler, R.W. (1967). Integrated pond systems for subdivisions. *Journal Water Pollution Control Federation*, 39, 1289–1298.

Pano, A. and Middlebrooks, E.J. (1982). Ammonia nitrogen removal in facultative wastewater stabilization ponds. *Journal Water Pollution Control Federation*, 54, 344–351.

Parker, C.D., Jones, H.L., Greene, N.C. (1959). Performance of large sewage lagoons at Melbourne, Australia. *Sewage and Industrial Wastes*, 31, 133–152.

Pearson, H.W., Mara, D.D., Bartone, C.R. (1987). Guidelines for the minimum evaluation of the performance of full-scale waste stabilization pond systems. *Water Research*, 21, 1067–1075.

Pearson, H.W., Avery, S.T., Mills, S.W., Njaggah, P., Odiambo, P. (1996). Performance of the phase II Dandora waste stabilisation ponds the largest in Africa: The case for anaerobic ponds. *Water Science and Technology*, 33, 91–98.

Pearson, H.W., Athayde, S.T. S., Athayde, G.B., Silva, S.A. (2005). Implications for physical design: The effect of depth on the performance of waste stabilization ponds. *Water Science and Technology*, 51, 69–74.

Racault, Y. and Boutin, C. (2004). Waste stabilisation ponds in France: State of the art and recent trends. In *6th International Specialised Conference on WSP*, Avignon.

Rajbhandari, B.K., Annachhatre, A.P., Vasel, J.L. (2007). Modeling of anaerobic treatment of was-tewater in ponds. *Water Science and Technology*, 55, 47–56.

Reed, S.C. (1985). Nitrogen removal in wastewater stabilization ponds. *Journal Water Pollution Control Federation*, 57, 39–45.

Reed, S.C. and Brown, D. (1995). Subsurface flow wetlands: A performance evaluation. *Water Environment Research*, 67, 244–248.

Reed, S.C. and Zirschky, J. (1988). The use of duckweed for wastewater treatment. *Research Journal of the Water Pollution Control Federation*, 60, 1253–1258.

Reed, S.C., Crites, R.W., Middlebrooks, E.J. (1995). *Natural Systems for Waste Management and Treatment*, 2nd edition. McGraw-Hill, New York.

Sah, L., Rousseau, D.P. L., Hooijmans, C.M., Lens, P.N. L. (2011). 3D model for a secondary facultative pond. *Ecological Modelling*, 222, 1592–1603.

Saqqar, M.M. and Pescod, M.B. (1992). Modeling coliform reduction in wastewater stabilization ponds. *Water Science and Technology*, 26, 1667–1677.

Saqqar, M.M. and Pescod, M.B. (1995). Modelling the performance of anaerobic wastewater stabilization ponds. *Water Science and Technology*, 31, 171–183.

Shaw, V.A., Meiring, P.G.J., van Eck, H. (1962). Preliminary results of research on raw sewage stabilization ponds. Report, CSTR, Pretoria.

Stratton, E. (1969). Nitrogen losses from alkaline water impoundments. *Journal of the Sanitary Engineering Division*, 95(2), 25–38.

Thirumurthi, D. (1974). Design criteria for waste stabilization ponds. *Journal Water Pollution Control Federation*, 46, 2094–2106.

Townshend, A.R. and Knoll H. (1987). Cold climate sewage lagoons. In *Proceedings of the 1985 Workshop*, Winnipeg.

Uhlmann, D., Recknagel, F., Sandring, G., Schwarz, S., Eckelmann, G. (1983). A new design procedure for waste stabilization ponds. *Journal Water Pollution Control Federation*, 55, 1252–1255.

Wehner, J.F. and Wilhelm, R.H. (1956). Boundary conditions of flow reactor. *Chemical Engineering Science*, 6, 89–93.

Zouhir, F. (2008). Modélisation d'une filière de traitement des eaux usées. Chenal algal à haut rendement. PhD Thesis, University of Liège, Liège.

3

Sludge Thickening

The biological treatment of wastewater generates sludge composed of organic and mineral matter, as well as various microorganisms. The density of this sludge is lower than that of primary sludge, and it is more difficult to dewater.

The goal is to reduce the sludge volume until it can be landfilled, incinerated or recovered in various forms.

The two main sources of sludge production are primary and secondary treatment of water. The solids settled during the first stage of primary and secondary settling are mixed with the sludge extracted during the sedimentation stage.

The sludge produced during the primary and secondary treatment of water by a wastewater treatment plant has the following characteristics:

– a high water content (98–99%), thus making it occupy a significant volume;

– a high concentration of organic matter that decomposes easily (putrefaction).

3.1. Origin and characteristics of sludge

3.1.1. *Primary sludge*

Primary sludge has not undergone biological treatment or decomposition. It is therefore highly unstable and putrescible. Approximately 60% of this sludge is considered suitable for settling. Its color is typically gray, and it has a high content of biodegradable solids and other types of waste. It readily releases its water content and thickens easily. Its water content varies between 98 and 99%. When chemical precipitation of phosphates occurs in the primary clarifier, iron hydroxides and iron phosphate give the primary sludge an ochre color.

3.1.2. *Biological sludge*

Biological sludge is commonly called "excess sludge" and comes from the activated sludge process. It is generally dark brown in color with an organic matter content that is partially decomposed.

Biological sludge has a damp earth smell, not unpleasant, but as it decomposes, it breaks down and produces foul odors. Its moisture content varies between 98 and 99%, and it is difficult to concentrate.

The benefits that can result from reducing the water content of sludge include:

– reduction in digester sizing requirements to achieve the same solid retention time;

– reduction in heat exchange capacity requirements for anaerobic digestion;

– reduction in sludge pumping and transport costs;

– reduction in ultimate disposal costs;

– reduction in handling problems and leachate production during sludge burial operations;

– better sizing for subsequent treatment or stabilization options (e.g. sludge drying, autothermal aerobic digestion) (ATAD)).

3.1.2.1. *Chemical conditioning of sludge*

Sludge thickening operations are highly dependent on sludge conditioning for their effective operation. Conditioning affects the solid concentration of the thickened sludge and the efficiency of solid capture.

Several characteristics of sludge, including industrially derived constituents, can hinder solid–liquid separation attempts. The presence of colloidal particles increases the specific strength of sludge and impairs sedimentation processes. The net negative charge exhibited by most sewage sludge tends to cause particles to repel each other, thus making them resist agglomeration into larger particles. The sludge particles have a bound water content which, if retained, results in a low solid content in the sludge cake after solid–liquid separation. Sludge conditioning operations attempt to modify one or more of the above sludge characteristics to improve the efficiency of the solid–liquid separation processes.

Sludge is generally treated using chemical methods, involving the addition of coagulants and/or polymers.

3.1.2.2. *Thickening types and sites within plants*

Sludge thickening can be used in the following locations within a treatment plant:

– before digestion for raw primary, secondary or mixed primary and secondary sludge;

– before dewatering;

– after digestion of the sludge or the supernatant;

– after the dewatering process to concentrate the filtrate, settled solids or centrate.

In most cases, the first step in sludge treatment is the thickening of biological and primary sludge, which involves initial dewatering before proceeding to more intensive treatment. Thickening is an intermediate dewatering process that reduces the size of the costly facilities required for further treatment.

Sludge thickening is the first stage of sludge volume reduction. This step occurs before digestion or any other sludge treatment. Sludge thickening typically refers to the process of reducing the free water content of the sludge, while dewatering refers to the reduction of the water content bound to flocs and capillaries in the sludge.

Three techniques are most often used for thickening:

– gravity thickening;

– flotation thickening;

– mechanical thickening.

3.2. Principle of sludge thickening

Thickening involves separating the interstitial water from the sludge particles. This step increases the sludge's retention time in the digester due to the resulting volume reduction.

It offers several advantages:

– further reduction of organic matter;

– increased dewatering yields;

– production of a buffer tank within the sludge treatment chain ensuring the homogeneity of the sludge physicochemical characteristics.

The sludge treatment processes in this stage allow for a reduction in the volume of sludge to be treated by removing water and thus increasing the solid concentration. The main objective is to increase the efficiency and economic optimization of subsequent processes.

The main thickening processes are:

– Gravity thickening: this process uses the force of gravity. The sludge is fed into the central zone, the thickened sludge is collected in the lower section and the supernatant remains in the upper section. This system is used for primary, physicochemical and mixed sludges that settle well using gravity. Biological sludge settles more slowly.

– Flotation thickening: the sludge concentrates in the upper part by the attachment of tiny air bubbles to the suspended solid particles, which are ultimately less dense than water. This type of system is suitable for thickening biological sludge due to its low settling capacity.

– Mechanical thickening: sludge concentration is achieved by increasing gravitational forces. Thickening tables or drainage tables, in which separation occurs through drainage of water via a moving horizontal porous belt, are suitable for activated or digested sludge. They are not suitable for physicochemical sludge.

3.3. Gravity thickening

This technique involves introducing the sludge suspension into a structure called a thickener, allowing the sludge to concentrate at the bottom. The sludge level must be such that it prevents any turbulence within the apparatus. The sludge retention time can range from a few hours to a few days. However, depending on the nature of the sludge, the retention time must be kept within a range that prevents sludge fermentation.

Sludge thickening by settling, or gravity thickening, is a process that takes place in a circular plant containing a slowly rotating part equipped with bottom scrapers, including a series of vertical spikes that facilitate the removal of volatile compounds by opening grooves in the sludge cover, thus facilitating thickening (see Figure 3.1). A bottom scraper sweeps the thickened sludge towards an outlet, from which it is periodically purged. A peak speed of 50–100 mm/s is required. A surface skimmer is recommended. A torque limiter is generally installed on the drive motor to prevent the rotary part from seizing and malfunctioning. The overall height of this equipment typically does not exceed 6 m, and the slope of the lower walls is approximately 15%.

Sludge origin	Concentration of thickened sludge (g.L^{-1})
Primary sludge + activated sludge	80–100
Primary sludge	100–150

Table 3.1. *Effectiveness of a gravitational thickener*

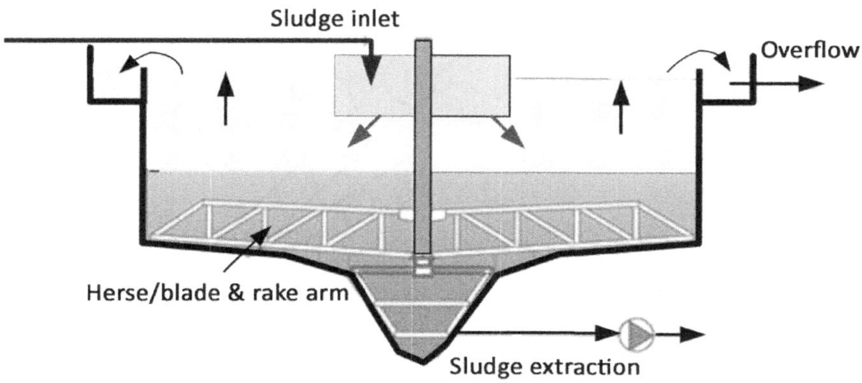

Figure 3.1. *Cylindrical thickener with scraping, side view*

Determining the size of a thickener has long required earlier laboratory analysis of the sludge. Indeed, the sludge settling curve in a test tube provides a first insight into the sludge settling suitability. The same phenomena observed during the settling of activated sludge are reproduced during thickening. The physical reactions that take place in the clarifier in the compaction zone are replicated in the thickener.

Thus, a sludge thickening curve has the same phases as a clarification curve, as described in Chapter 3 of volume 3 of this work. The Coe–Clevenger and Kynch theories aid in determining the thickener height. Once concentrated, the sludge is removed from the unit and fed into the other stages of the treatment process. This type of system incorporates mechanization that allows for recycling and slow agitation of the sludge.

The advantage of such a system lies in the ease with which the thickened sludge slides towards the central pit, from which it is extracted. Interstitial water is removed using a vertical harrow connected to the bottom scraper and the surface skimmer.

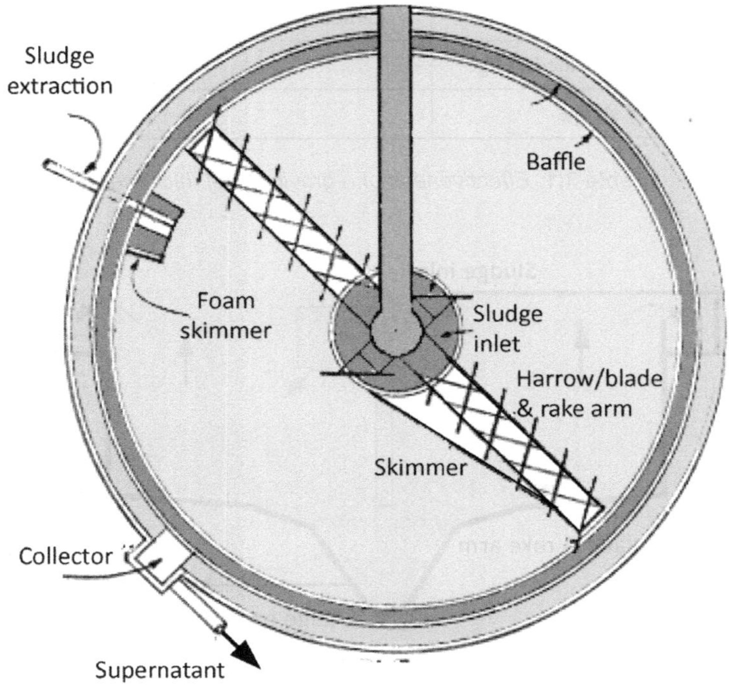

Figure 3.2. *Cylindrical thickener with scraper, top view*

3.3.1. *Thickener surface area*

From the settling curve, it is possible to determine the heights and concentrations of the sludge after settling times of 1, 2, 6 and 24 hours. The initial settling velocity, given by the tangent to the initial part of the curve, is also calculated, as is the surface area of the structure required to thicken a given sludge flow rate Qb ($m^3.h^{-1}$) of an initial concentration Xo ($g.L^{-1}$) towards a final concentration Xb ($g.L^{-1}$) for a total height of the structure H. Using the Kynch method, the settling curves allow the following calculations:

– initial settling velocities ($m.h^{-1}$);

– applied charges ($kg.m^2.d^{-1}$);

– thickener surface area.

3.3.2. *Determining the applied load*

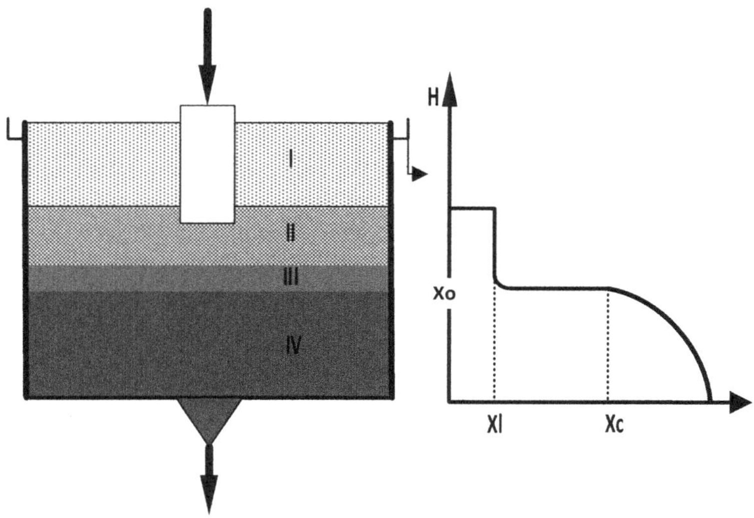

Figure 3.3. *Sludge stratification and concentrations. For a color version of this figure, see www.iste.co.uk/gaid/wastewaterengineering4.zip*

The sizing of a mechanical thickener is based on the sludge settling curve according to Kynch's theory, which states that the settling velocity of a given particle type depends only on its local concentration. The height of the supernatant (zone I, Figure 3.3) is maintained at a minimum of 0.5–1 m. Flocculant addition allows for safe control of the effluent quality. When zone I is invaded by solid particles, the thickener is said to be overflowing.

Below this zone lies zone II, called the halted settling zone. It contains sludge in a uniform concentration, which settles at a constant rate. This zone can have a concentration between the inlet concentration (Xc) and a limiting intermediate concentration (Xl). Under normal operating conditions, the concentration (Xl) of zone II depends on the sludge feed rate rather than the initial sludge concentration. Xl is low if the feed rate is low and increases with the feed rate, reaching a limit when solids settle at the maximum possible rate in this zone. If solids are introduced at a flow rate exceeding this maximum, the concentration in zone II will continue to be linked to the maximum feed rate. In this case, excess solids will not be deposited through zone II but will pass into zone I and be removed via the overflow.

Below zone II, a region characterized by a concentration gradient is created, zone III, called the transition zone because it marks the transition from a sludge of constant concentration to the solids located below it.

This transition zone is fragile and could end up being divided between zone II and zone IV.

At the bottom of the thickener is zone IV, called the compression zone, which contains compacted sludge. In this region, the flocs lie directly on top of each other, with those at the top exerting pressure on those below, thus creating a concentration gradient. This zone contains equipment such as the harrow and scraper, whose circular motion further concentrates the sludge, inducing an additional concentration gradient.

The concentration of the extracted sludge therefore corresponds with the compaction zone. The greater the height of accumulated sludge, the higher the concentration.

The height of each zone in the thickener depends on the settling characteristics of the sludge. The treatment capacity of each zone is defined as the quantity of solids that passes from one zone to another per unit area and per unit time.

It is estimated that the settling rate is a function of the sludge concentration. If no suspended solids leak through the overflow, the total solid flow must pass through all zones of the thickener under steady-state conditions, so zones with a smaller treatment capacity will have a greater depth at the expense of zones with higher treatment capacities.

A thickener is said to overflow if solid particles pass from zone II to zone I. There are three reasons for this:

– The load contains very fine particles that cannot settle.

– The loading flow rate exceeds the settling capacity of zone II.

– The feed rate exceeds the removal rate. In this case, solid particles accumulate in the thickener and are eventually transported to zone I.

Kynch's formula provides a graphical construction from which the unit area of the thickener is determined to treat the sludge mass (solids) per unit time. A simple relationship links the initial concentration and interface height with other concentrations and heights along the structure:

$$Ho\ Xo = H\ X = Hb\ Xb$$

where:

– Ho: initial height of the mixed liquid (m);

– Xo: initial sludge concentration ($g.L^{-1}$);

– Hb: height corresponding to the concentration in Xb (m);

– Xb: sludge concentration at the bottom of the structure ($g.L^{-1}$).

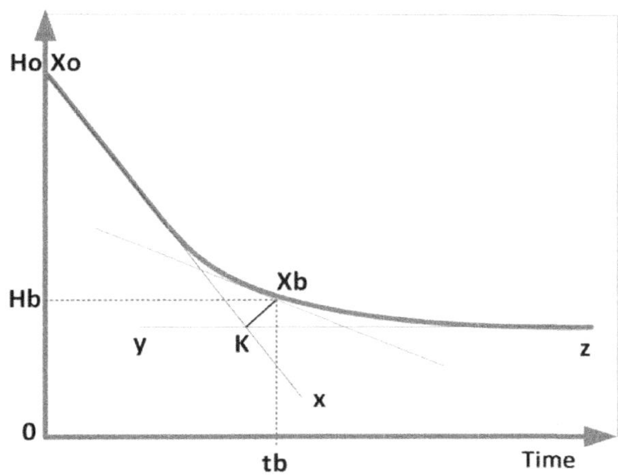

Figure 3.4. *Determining the thickened sludge concentration. For a color version of this figure, see www.iste.co.uk/gaid/wastewaterengineering4.zip*

The procedure is as follows:

1) Plot the height of the settled sludge in the test tube as a function of time and represent the slope of the line on the curve (see Figure 3.4). From this, the settling (or thickening) rate can be deduced in $m.h^{-1}$.

2) Next, construct the slopes Hx and yz of the settling curve. They intersect at point K.

3) Construct the bisector of angle HKZ. Xb is the intersection point of the bisector and the curve.

4) Draw a tangent at point Xb (which is the sludge compaction point); using the relation HoXo = Hb Xb, it is possible to determine the sludge height (Hb), the applied load and the thickener area.

5) Knowing Xb, we can deduce the corresponding height:

Hb = HoXo/Xb

6) Read the settling time corresponding to Xb and Hb.

7) The load applied to the thickener is calculated as follows:

TSS load applied ($kg.m^2.d^{-1}$) = Ho Xo/tb

where Ho is in m, Xo is in $kg.m^{-3}$ and tc is in d.

The required surface area is therefore calculated as follows:

Surface area required (m^2) = Qb Xo (kg $TSS.d^{-1}$)/applied load ($kg.m^2.d^{-1}$)

EXAMPLE.–

Table 3.2 shows the calculation of the required surface area for an existing station.

Parameters	Formulas	Values
Sludge concentration at the inlet ($kg.m^{-3}$)	Xo	10
Inlet sludge flow rate ($m^3.d^{-1}$)	Qb	15,120
Thickness height (m)	Ho	6.2
Sludge concentration at the outlet ($kg.m^{-3}$)	Xb	40
Compacted sludge height (m)	Hb = HoXo/Xb	1.55
Thickening time (h)	tb	3.5
Inlet loading rate ($kg.d^{-1}$)	Ch_b	151,200
Applied loading rate ($kg.m^{-2}.d^{-1}$)	24HoXo/tb	425
Total surface area required (m^2)	Inlet load/applied load	356
Number of thickeners	n	2
Type	–	Cylindrical
Unit surface area (m^2)	S total/2	178
Unit diameter (m)	Du	15.05
Unit volume (m^3)	Vu	1,103

Table 3.2. *Calculation example of the thickener surface area*

The following relationship is given to determine the thickening surface area (Edde and Eckenfelder 1967):

$$S\,(m^2) = \frac{t\,QbXo}{HoXo} = \frac{t\,QbXo}{HbXb}$$

where t is in hours, Qb is in $m^3.h^{-1}$, X is in $kg.m^{-3}$ and H is in m.

3.3.3. *Mass loading rate ($kg.m^{-2}.d^{-1}$)*

In a mechanical sludge thickener, the specific loading is highly variable. It is expressed in kg $MS.m^{-2}.d^{-1}$. Throughout the process, a thick layer of sludge remains at the bottom to facilitate better concentration. An operational variable for this equipment is the sludge volume ratio (SVR), which is the volume of the thickener sludge layer divided by the volume of the sludge thickened daily. SVR values range from 0.5 to 2 days, with the lowest values observed in warm weather.

The variables of a thickener can be linked by the relation (Edde and Eckenfelder 1967):

$$\frac{Xb}{Xo} - 1 = D\,C'm^{-n}$$

where Xo is the concentration of solids at the thickener inlet ($kg.m^{-3}$); C′m is the applied mass loading rate ($kg.m^{-2}.d^{-1}$); D and n are constants related to the sludge properties, the sludge inlet concentration and the test conditions; n is 0.25 for urban sludge (primary activated); and D can vary between 10 and 90.

A simple model associating sludge velocity and concentration has been proposed for sizing thickeners (Coe and Clevenger 1982):

Applied loading rate = v/(1/Xo – 1/Xb)

with:

– load: flow of solids per unit area and time ($kg.m^{-2}.\ h^{-1}$);

– v: velocity relative to the overflow rate ($m.h^{-1}$);

– Xo: initial TSS concentration ($kg.m^{-3}$);

– Xb: concentration of the thickened sludge ($kg.m^{-3}$).

This relationship underestimates the thickener area because the velocity parameter must be small to avoid turbulence and TSS leakage into the overflow. To recover the previously calculated area, the surface velocity must be equal to 1.3 $m.h^{-1}$, which is high for a conventional gravity thickener. With the values of Xo and Xb respectively equal to 10 $kg.m^{-3}$ and 40 $kg.m^{-3}$, the required surface area in the previous example is obtained from the applied loading rate.

Thus:

Applied loading rate = 1.3 $m.h^{-1}$/(1/10 – 1/40) = 17.33 $kg.m^{-2}.h^{-1}$ = 415 $kg.m^{-2}.d^{-1}$

For a sludge loading rate of 6,300 $kg.h^{-1}$ (i.e. 151,200 $kg.d^{-1}$ in the previous example), the required surface area is therefore:

S = 6,300/17.33 = 363.5 m^2

3.3.4. *Total height of a gravity thickener*

As in a clarifier, the total height of a thickener is the sum of different heights, including:

– the sludge settling zone;

– the halted settling (thickening) zone;

– the constant-speed sludge thickening zone;

– the compaction zone.

A total height of 3–6 m is generally accepted, with a minimum of 3 m for most sludges. The overflow water height is between 1 and 2 m.

There are two main types of gravity thickeners, depending on whether they are mechanized or not:

– A static thickener, with a conical bottom, has a diameter of less than 5 m; the base slab is angled at 45–60° to accelerate settling.

– A mechanical thickener has a flat bottom, slightly inclined relative to the horizontal (10–15°). It is equipped with a scraper to facilitate the movement of sludge towards the central pit, and also to facilitate the removal of interstitial water using a harrow (see Figure 3.5). The diameter of this structure exceeds 5 m and often reaches 40 m. The power of the scraper is approximately 1 $W.m^{-3}$.

3.4. Lamellar thickening

Here, two functions are combined:

– gravity thickening, for which the load on the foundation slab and the sludge height are predominant;

– clarification in the lamellar zone, for which the Hazen velocity is the limiting parameter.

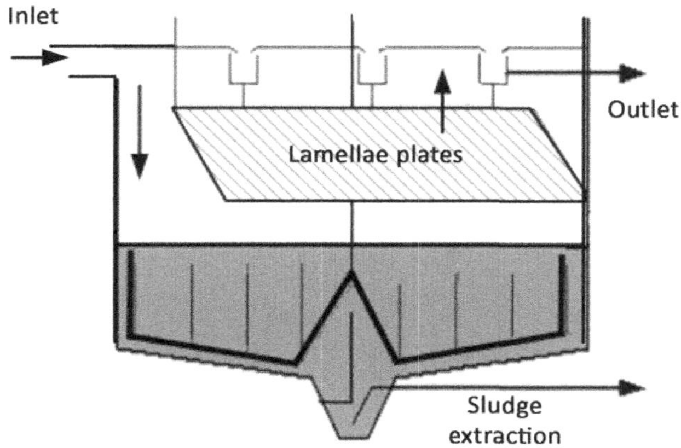

Figure 3.5. *Lamellar thickening without reagent injection*

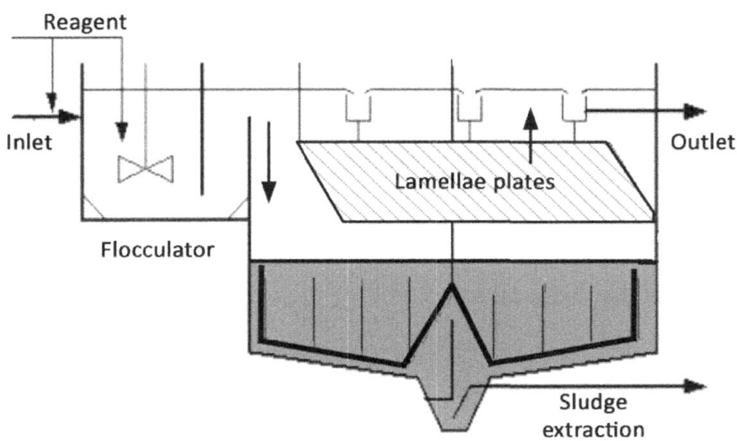

Figure 3.6. *Lamellar thickener with reagent injection*

A lamellar thickener (see Figures 3.5 and 3.6) involves:

– a feeding zone across the entire width of the settling tank;

– removal of floating debris by surface sweeping or raising the water level (dumping);

– an equidistribution grid;

– sludge extraction.

Figure 3.7 shows two configurations, one with a frontal sludge supply, the other with a lateral supply.

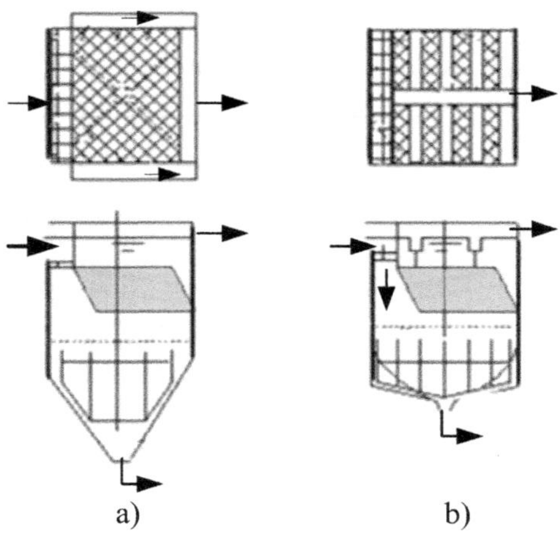

Figure 3.7. *Thickeners with two types of sludge supply: a) frontal; b) lateral*

The main factors that influence thickening are:

– optional flocculation;

– the surface area of the foundation slab;

– the depth of the sludge layer;

– harrowing;

– the sludge extraction method.

The thickening zone is located below the supply and settling zones.

3.4.1. *Thickener surface area*

This is an essential parameter of any static thickener, whether conventional or lamellar. The surface area is defined by the maximum flow of material allowed per day and per square meter of structure (base load).

The reference surface area taken into account for hopper-type thickeners and circular scrapers is the total (square) surface area of the device. When the total surface area required for thickening exceeds that determined by the base load, it is this surface area that determines the size of the structure.

3.4.2. *Sludge layer height*

A sludge layer 3–4 m above the extraction point (bottom of the conical or pyramidal hopper) is recommended to ensure optimal concentration. Two sludge bed detection probes are installed to ensure the proper functioning of this system.

A height of 1–1.5 m is maintained between the plate support structure and the sludge bed, with the total distance between the support structure and the bottom being 4.5–5 m.

3.4.3. *Retention time*

A significant base slab load allows for a substantial reduction in retention time. For example, a 4 m layer of sludge at 20 $g.L^{-1}$ is renewed every one to two hours when the average incoming hourly load is 1,500 kg $MS.m^{-2}.d^{-1}$.

3.4.4. *Harrowing*

The continuous movement of a harrow in the sludge bed helps remove water and gas pockets, homogenize the different layers and remove short circuits at the suction point. The length of the harrow bars must ensure that the sludge layers are turned over to a minimum height of 3 m.

3.4.5. *Removal of thickened sludge*

A high-flow purge-and-trap ensures a compact sludge layer, a stable water/ sludge interface, good clarification and clean lamellae.

3.4.6. *Storage of thickened sludge*

Sufficient storage capacity will compensate for the gap between upstream production and downstream processing and will ensure proper operation of the thickener.

3.4.7. *Discharge of clarified water*

The concentration of TSS in the supernatant depends on the size of the structure and the operating conditions. A significant improvement in the quality of the discharged effluent can be achieved by introducing a polymer at the sludge inlet. For this purpose, a mixing zone of a few minutes is necessary for the formation of large flocs. This can vary between 0.5 and 2 mg.L^{-1}.

3.4.8. *Laminar cross-sectional layout*

Settling is ensured by a counter-current lamellar system fed from below.

Parameters	**Type of lamellae**	
	Inox (flat plates)	**PVC (alveolar structure)**
Inclination (°)	60	60
Average spacing e (mm)	100	80
Plate length L (m)	1.5	1.5
L/e ratio	15	18.7
Plate width (m)	1.03	1.22
Number of plates per module	5	8
Water length (m)	0.577	0.74
Water surface area* (Sm, m^2)	0.595	0.903
Total projected area * (TP, m^2)	3.862	7.358
TPS/Sm*	6.5	8.15
Module volume V (m^3)	0.773	1.173
Dry weight** (kg)	60	58.7
Weight/volume** (kg.m^{-3})	77.5	50
Weight/TPS ** (kg.m^2)	15.5	8

* Values related to plate dimensions; ** estimated values, depending on suppliers. TPS: total projected surface; Sm: mirror surface.

Table 3.3. *Main characteristics of the plates used for sludge thickening*

To prevent clogging resulting from accidental sludge infiltration into the plates, the following is advisable:

– a minimum spacing of 80 mm between the slats;

– a simplified support structure;

– the absence of inter-plate fixings and limited dead spaces;

– a small water surface area occupied by the gutters.

The main dimensions are summarized in Table 3.3.

Table 3.4 presents an example of calculations for Hazen velocities of 0.5–1 $m.h^{-1}$.

Parameters	**Inox (flat plates)**		**PVC (alveolar structure)**	
Average spacing (mm)	100		80	
Plate length (m)	1.5		1.5	
Hydraulic diameter (m)	0.182		0.109	
Hazen velocity ($m.h^{-1}$)	1	0.5	1	0.5
Surface velocity ($m.h^{-1}$)	6.5	3.25	8.15	4.07
Circulation rate ($m.h^{-1}$)	7.5	3.75	9.4	4.7
Reynolds number	330	165	248	124

Table 3.4. *Hazen velocities and surface velocities for both types of lamellar supports*

The sizing procedure includes the calculation of:

– the total projected surface area based on the flow rate to be treated and the desired Hazen velocity;

– the surface area corresponding to the type and height of the chosen slats (according to the ratio TPS/Sm);

– the size and number of modules needed to fill the calculated mirror surface as much as possible.

3.5. Flotation thickening

Flotation is a process for thickening sludge primarily resulting from biological treatment, either activated sludge (settled or not) or biofilter wash water. This process can also be applied to some tertiary sludge.

3.5.1. *Principle*

Dissolved air flotation (DAF) allows for the separation of solid particles and liquids, as well as the separation and thickening of sludge. It works by reducing the apparent mass of the sludge through the adsorption of fine gas bubbles, causing them to rise to the surface. Solid particles attach to these very fine air bubbles, giving them a very rapid upward velocity. The air bubbles are formed by dissolving air in water at high pressure, then depressurizing and releasing it into a flotation tank. The sludge rises at a much higher rate than with gravity sedimentation. The thickened sludge is collected from the surface using a scraping system. DAF systems achieve very high performance in both municipal and industrial wastewater treatment plants.

The implementation of this thickening method allows:

– larger specific loads;

– smaller tank dimensions;

– thicker sludge than with gravity thickening;

– greater removal of fine particles and colloids than with gravity thickening (without reagents).

With the injection of reagents (coagulant/flocculant), flotation tanks occupy approximately five times less space than static settling tanks for the same treated flow rate. Furthermore, they can be used to thicken the already separated sludge and generally allow for an increase in sludge concentration of up to 8–10% dry matter, depending on the sludge origin.

Two methods of promoting flotation using different gases and applications are dispersed air flotation and DAF. The latter method is the most widely used. It consists of simply injecting pressurized air or mixing and expanding a fraction of the separated interstitial water, which is then recycled after saturation with compressed air. The main variables in such a system are the recycling rate, the gas pressure, the air/solids ratio and the composition of the interstitial water. The bubbles retained for pressurization are approximately 30–80 microns in diameter.

The following values are recommended:

– mass loading rate: 50–120 kg TSS.m^{-2} d^{-1};

– hydraulic loading rate: 5 m.h^{-1};

– flotation time: 10–40 min;

– float blanket thickness: 20–50 cm;

– air consumption: 0.03–0.07 kg.kg^{-1} TSS.

As with gravity thickening or sedimentation, Stokes's law applies, but in a force field opposite to gravity.

The basic principle involves dissolving air in water in a pressure tank. This mixture is then introduced and expanded to ambient pressure in the flotation tank using a suitable device. The expansion, combined with strong turbulence, causes the formation of a cloud of fine bubbles, 30–80 µm in diameter, which must be distributed as evenly as possible within the active section of the flotation tank.

As these fine bubbles rise, they adhere to the suspended solids (forming aggregates with a density of approximately 0.6–0.7) and carry them to the surface of the tank for collection (skimming). Simultaneously, the clarified water is drawn off from the bottom (underflow) of the flotation tank.

3.5.2. *Direct flotation*

The raw effluent is totally or partially pressurized in a pressurized vessel using air.

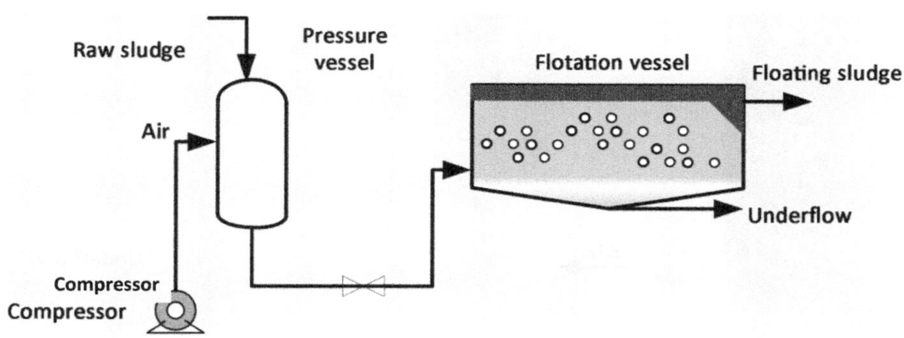

Figure 3.8. *Direct flotation*

3.5.3. *Indirect flotation*

In this case, clarified wastewater or filtered water is brought into contact with pressurized air. After expansion, this water forms "white water", similar to water coming out of a pressurized tap. This type of flotation device therefore requires a recirculation loop. The raw sludge can either enter the white water recirculation loop, in which case it is called co-current flow, or be injected into a mixing tank located in the tank where the recirculated white water enters, in which case it is called counter-current flow, as described in Figure 3.10.

Floatation tanks are often of the indirect, counter-current type, meaning they mix white water with sludge. Unlike direct ones, this design allows for better capture efficiency (the underflow usually contains less than 100 mg.L^{-1} of TSS) at the cost of higher energy consumption due to the power of the recirculation pump.

The expansion of the pressurized fluid therefore occurs through pressure loss via strainers.

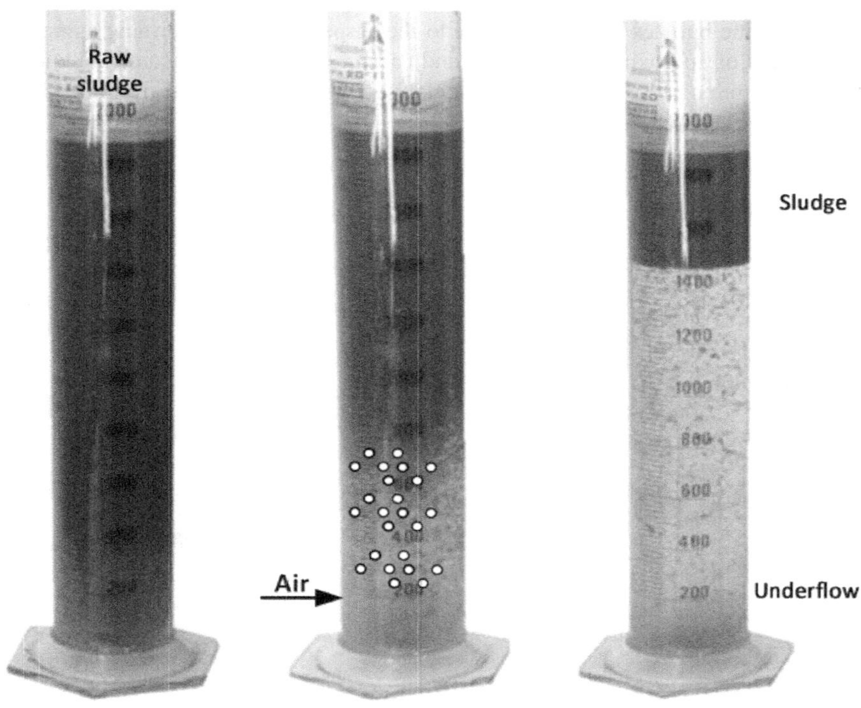

Figure 3.9. *Principle of sludge flotation*

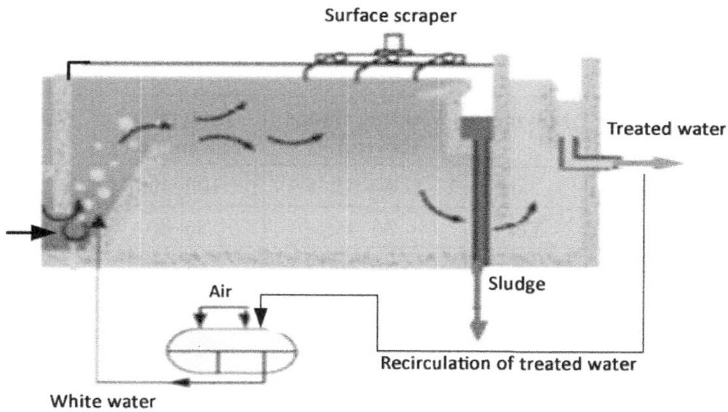

Figure 3.10. *Indirect dissolved air flotation*

Many flotation vessels include the following zones:

– a contact zone where the sludge and desaturated water (white water) are mixed;

– a transition zone between the floating sludge and the clear water;

– a sludge compaction zone where a surface scraper collects the floating sludge;

– a hydraulically stabilized clarification zone where the underflow is recovered via perforated piping;

– a settling zone allowing the tank to be purged of dense elements using a bottom valve.

Taking these five zones into account allows for the calculation of the vessel height. The pressure tank enables rapid gas–liquid transfer.

For all types of thickeners, performance is linked to the nature of the sludge being treated, the two main applications of which are:

– activated sludge flotation at the clarifier outlet;

– flotation of biofilter wash water.

The presence of polymers therefore primarily affects the quality of the underflows and consequently the capture rate. It should be noted that a too high quantity of polymer can result in a significant degradation of performance due to the increased viscosity of the recirculated float water. A high recirculation rate combined with polymer treatment can, in some cases, achieve underflows with a lower titration of

30 mg.L^{-1} (95th percentile) and thus avoid recirculating the resources at the beginning of the biological treatment.

Sludge type	Polymer	Floating sludge TSS (g.L^{-1})	Underflow TSS (mg.L^{-1})	Capture rate (%)
Activated medium load	No	30–35	100–400	> 80%
	Yes	35–45	50–80	> 95%
Activated low load	Yes	40–50	15–50	> 98%
Secondary biofilter	No	30–40	100–200	85–90%

Table 3.5. *Indirect flotation efficiency*

Figure 3.11. *Floating sludge*

The following design parameters are given for guidance purposes only, to indicate the normal range of values encountered in large-scale operation:

– Mass load: 48 kg.m^{-2}.d^{-1} (without flocculating chemicals) and up to 240 kg.m^{-2}.d^{-1} (with flocculating chemicals).

– Most facilities now use chemical conditioning with polymers to achieve a more economical operation.

– Polymer injection: 0–25 $g.kg^{-1}$ DS (DS: dried solids).

– Hydraulic loading rate: up to 5 $m^3.m^{-2}.h^{-1}$, based on the total throughput including recycling, when polymers are used.

– Without chemicals, a lower hydraulic charge is recommended.

– Retention time is not critical provided that the particle upwelling rate is sufficient and that the horizontal velocity in the unit does not produce turbulence in the sludge blanket.

– Removal of thickened sludge: a surface scraper moves the thickened sludge towards a sludge hopper.

– Bottom sludge: a collector for settled sludge is attached to the base of the structure. This waste is discharged into a hopper adjacent to the flotation unit.

These performance levels are achieved by adhering to the conditions defined in Table 3.6.

	Activated sludge	**Biofilters**
Concentration at the inlet	2–12 $g.L^{-1}$	0.2–2 $g.L^{-1}$
SVI	150–450 $ml.g^{-1}$	50–100 $ml.g^{-1}$
Polymer-free mass load	$<$ 2.5 kg $DS.m^{-2}.h^{-1}$	–
Polymer mass loading	$\leq$ 10 kg $DS.m^{-2}.h^{-1}$	$\leq$ 8 kg $DS.m^{-2}.h^{-1}$
Hydraulic loading rate	$<$ 5 $m.h^{-1}$	$<$ 5 $m.h^{-1}$
Anionic polymer	1–2 $g.m^{-3}$	2–3 $g.m^{-3}$
Energy consumption/m^3 of water	0.3–1.2 kWh/m^3 treated	0.05–0.5 kWh/m^3 treated
Energy consumption/ton	50–150 kWh/T DS	100–600 kWh/T DS
Relative pressure of the pressurization vessel	5–6 bar	5–6 bar

Table 3.6. *Floatation vessel sizing parameters*

The thickness of the sludge layer to be maintained depends on the amount of floating matter removed. It is generally between 50 and 100 cm.

3.5.4. *Determining the size of a flotation vessel*

3.5.4.1. *Data*

The nature of the sludge is assessed primarily by its sludge volume index (SVI). The inlet concentration is Xe and the flow rate is Qe.

3.5.4.2. *Calculations of loading rates and main geometric parameters*

Knowing the sludge load to be treated, the concentration of the floating sludge XF, with or without polymer, is determined at the chosen mass loading Cm. At a given recirculation rate τR, the capture efficiency Tc, calculated from the TSS, depends on the inlet sludge concentration Xe.

3.5.4.3. *Surface parameters*

The equipment surface area is determined based on parameters such as solid loading rate and surface loading rate (also known as upflow velocity). The solid loading rate is expressed in kg $TSS.m^{-2}.d^{-1}$ and the surface loading rate in $m^3.m^2.h^{-1}$.

The surface area of the base slab is given by:

$$Sr = \frac{QeXe}{Cm}$$

– Volume and depth of structure: for a predefined height (H), the volume is calculated by:

$$V = H.Sr$$

– Hydraulic loading rate:

$$Ch = \frac{Cm}{Xe(R+1)} \text{ with } Ch < 5 \text{ m.h}^{-1}$$

– Underflow TSS concentration:

$$Xs = \frac{(QeXe - QfXf)}{Qs}$$

where:

- Xs: TSS concentration of the floating water (underflow) ($kg.m^{-3}$);
- Xf: TSS concentration of the floating sludge ($kg.m^{-3}$);
- Xe: TSS concentration of the inlet sludge ($kg.m^{-3}$);

- Qe: sludge inlet flow rate ($m^3.h^{-1}$);
- Qf: floating sludge flow rate ($m^3.h^{-1}$);
- Qs: floating water flow rate ($m^3.h^{-1}$).

3.5.5. *Operating parameters*

For a temperature T in °C and the atmospheric pressure of the gaseous atmosphere Pa expressed in atmospheres, Henry's law gives the equilibrium mole fraction y_a of air in pure water, expressed in mol/mol:

$$y_a = Pa/K_H\,(T)$$

$K_H\,(T)$ being the Henry pseudo-constant given in Figure 3.12.

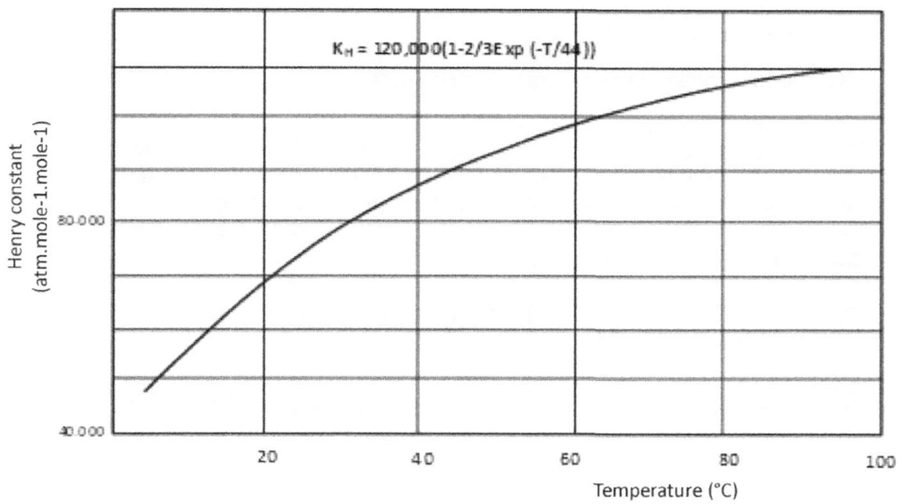

Figure 3.12. *Henry constant (K_H) of air in water depending on the temperature*

Using the same units, we can immediately deduce the air concentration Ca in $kg.m^{-3}$ (or $g.L^{-1}$):

$$Ca = 1{,}000\,\frac{28.8}{18}\,y_a = 1{,}600\,y_a$$

The air/sludge ratio As can be obtained using the equation: $As = K.Xe^{-0.36}$.

K is between 0.04 and 0.06 if A is in kg air/kg MS and Xe in $g.L^{-1}$.

The sludge height Hb depends linearly on the floating sludge concentration Xf according to the equation:

$$Hb = a\ Xf + b$$

If Xf is > 20 $g.L^{-1}$, Hb can be estimated using the values of a = 0.06 m/g/L and b = 1.40 m.

3.5.6. *Modeling the efficiency of a flotation vessel as a function of load*

Assuming a limitation in mass load Cm only, we can write the following.

3.5.6.1. *Concentration of floating sludge*

The parameters are as follows:

– Xf, the floating sludge concentration ($kg.m^{-3}$);

– Xe, the inlet sludge concentration ($kg.m^{-3}$);

– Xf max, the maximum achievable floating sludge concentration ($kg.m^{-3}$), which is inversely related to the SVI and increases with polymer dosage.

This model is written as follows:

$$\mathrm{Xf} = (\mathrm{Xfmax} - \mathrm{Xe})\left(1 - \frac{\mathrm{Cm}^{\mathrm{n}}}{\mathrm{Cm}^{\mathrm{n}} + \mathrm{b}}\right) + \mathrm{Xe}$$

with n and b being the constants of the experimental model. The load Cm is expressed in $kg/h/m^2$.

When the load becomes very large, the output Xf tends towards the input concentration Xe.

The value of Xf max can be approximated by the relation:

$$Xf\ max = 83\ Exp\ (-0.0037\ SVI)$$

3.5.6.2. *Capture rate*

The capture rate η can be approximated by the following smoothing formula:

$$\eta = 0.99 - EXP\ (-\lambda\ (1 - EXP(Xe/800)))$$

Xe, the inlet TSS concentration, is expressed in $mg.L^{-1}$. λ varies between 3 and 6.

3.5.6.3. *Underflow concentration*

The underflor concentration is obtained from the capture rate η.

– Xsmax, the underflow outlet maximum concentration;

– Xsmin, the underflow outlet minimum concentration:

$$Xs = (Xsmax - Xsmin)\left(1 - \frac{Cm^n}{Cm^n + b}\right) + Xsmin$$

with n and b being the constants of the experimental model. The charge Cm is expressed in $kg.h^{-1}.m^{-2}$.

Xsmax is approximated by the relation:

$Xsmax = 1{,}200/SVI$ or $1{,}968\ SVI^{1.106}$

3.6. Comparison of conventional gravity thickeners and flotation thickeners

Table 3.7 shows the expected performance for the two thickening systems.

Thickening process	Sludge type	Expected performance
Gravity	Primary sludge	8–10% DS
Gravity	Primary and biological sludge	5–8% DS
Gravity	Biological sludge	2–3% DS
Gravity	Digested primary sludge and biological sludge	8–14% DS
Flotation (DAF)	Biological sludge	4–12% DS and ≥ 95% capture rate with chemical reagents

Table 3.7. *Expected performance for the two thickening systems*

Flotation is only recommended for biological sludge that is less dense than primary sludge. Depending on the type of sludge, gravity thickening leads to dry solid contents varying between 2% and 14%.

The advantages of DAF over gravity thickeners for biological sludge include its reliability, the production of higher sludge concentrations and better solid capture. Its disadvantages include the need for greater operational expertise and higher operating costs.

Table 3.8 summarizes the advantages and disadvantages of the two systems.

Dissolved air flotation (DAF) thickener	Gravity thickener
Advantages	
Thickening efficiency: DAF achieves a high concentration of suspended solids in the water by forming microbubbles of air. This significantly reduces sludge volume. The concentration obtained in the thickened sludge is in the range of 4–6%, with mixtures of primary and activated sludge in excess (even ≥ 8%). The use of polyelectrolytes as sludge flocculants improves solid separation from 85% to 98–99%. Rapid separation: flotation allows for rapid separation of solids, which is beneficial in situations with high sludge loads. Less space required: DAF occupies less space than gravity thickeners.	Low initial costs: gravity thickeners are less expensive in terms of initial investment. Reduced energy consumption: no ventilation system is required, which reduces operating costs. Robustness: they are simple and robust.
Disadvantages	
Initial costs: microbubble air injection technology and recirculation systems can increase initial costs. Energy consumption: microbubble production and sludge recirculation require more electrical energy.	Thickening rate: the sludge velocity is significantly lower than that of the DAF. Larger footprint: they take up more floor space. Lower efficiency: the concentration of solids is not as high as in the DAF.

Table 3.8. *Technical comparison of the main sludge thickening systems*

3.7. Dewatering table

The dewatering table is designed for thickening industrial or municipal sewage sludge. Thickening is an essential step in reducing the volume of sludge from wastewater treatment plants. It optimizes the subsequent conditioning, stabilization and dewatering phases by ensuring optimal separation of liquid and solids. The dewatering table thickens the sludge by increasing the amount of dry matter, thus minimizing the volume of water.

The thickening process begins with flocculation of the sludge. This first step is particularly important for allowing the dry matter to agglomerate into "flocs", providing an initial liquid/solid separation. This step involves polymer injection into the sludge. Generally, the flocculant is injected upstream of the sludge pump to benefit from the progressive mixing achieved by the screw conveyor feeding the pump.

This step aims to give the sludge a "structure" so that it can effectively undergo the subsequent stages. Thus, as a general rule, a dewatering table cannot function without polymer injection.

The sludge to be thickened is mixed with a flocculant polymer and then pressed onto a 600 μm mesh screen. The flocculated sludge flows uniformly onto the filter belt. The water released during flocculation then flows freely through the mesh of the filter belt. Without prior conditioning, the sludge would simply spread out on the belt and flow down the sides.

The sludge flows over a perforated tray through which some of the water can drain by gravity (see Figure 3.13). This step is often preceded by the addition of flocculants to facilitate the separation of the solid and liquid phases of the sludge.

The filtrate is collected and then generally returned to the headworks of the treatment plant via a septic tank or treated in another way. To prevent clogging of the filters, injection nozzles allow them to be cleaned with water (industrial or potable). The thickening achieved by this system results in a dry matter content of 4–10% with a capture rate of 96% or higher and low polymer consumption.

The dewatering tables consist of a stainless steel frame, a sludge feeding and distribution system, a gravity thickening zone, a pressing zone, a geared motor, a continuous filter washing zone, filter tensioning and/or tracking systems, filtrate recovery tanks, integrated piping and electrical wiring.

This compact and highly flexible technique allows for a reduction in sludge volume by a factor of approximately 6–7, depending on the sludge characteristics.

The sizing parameters include sludge age, retention time, storage tank size, inlet concentration, MVS/TSS ratio and polymer dose.

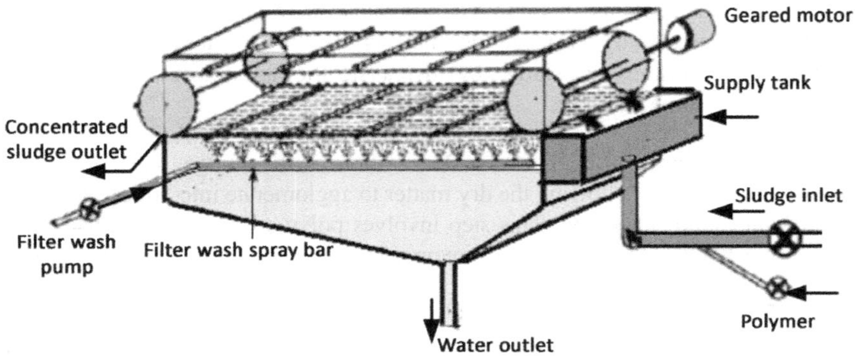

Figure 3.13. *How a draining table works*

Figure 3.14. *Draining table*

3.8. Conclusion

Thickening plays a role in all sludge treatment and disposal schemes. Even if this step is not used in a specific plant, thickening will always be involved in clarifiers. Thickening has a significant impact on treatment costs and the additional steps required.

The type of sludge to concentrate in a thickener depends on factors such as the nature of the sludge, the size of the treatment plant and other treatments involved in the sludge process. These factors offer significant potential for cost reduction.

Gravity thickening is characterized by its low energy consumption, while flotation thickening requires higher operating costs.

Nevertheless, a well-calculated thickening process provides enormous benefits to the operator, as properly designed waste treatment results in safer and more consistent operations. Instrumentation should be considered for all thickening processes to monitor the flow of contaminants and effluents, and sampling should confirm the presence of contaminants and thickened solids.

3.9. References

Bueno-de-las-Heras, J.L., Gutiérrez-Lavín, A., Mahamud-López, M.M. (2024). Applying a novel generalized model of sedimentation to the design of settlers and thickeners: Integration and extension of the classical procedures. *Ann. Environ. Sci. Toxicol.*, 8(1), 043–056.

Canler, J.-P., Cauchi, A., Duchène, P., Fernandes, P., Larigauderie, A., Leboucher, G., Pujol, R. (2002). L'épaississement des boues : les règles de bonne gestion. *Tech. Sci. Méthodes*, 4, 55–68.

Coe, H.S. and Clevenger, G.H. (1916). Methods for determining the capacities of slime-settling tanks. *Trans. AIME*, 55, 356.

Coe, H.S. and Clevenger, G.H. (1982). Determination of thickening unit areas. *Trans. AIME*, 55(3), 28–39.

Cole, R.F. (1968). Experimental evaluation of the Kynch theory. PhD Thesis, University of North Carolina, Chapel Hill.

Crittenden, J.C., Rhodes Trussell, R., Hand, D.W., Howe, K.J., Tchobanoglous, G. (2012). *Water Treatment Principles and Design*, 3rd edition. John Wiley & Sons, New York.

Davis, M.L. (2010). *Water and Wastewater Engineering: Design Principles and Practice.* McGraw-Hill, New York.

Davy, W. (2016). Etude de la filière boue d'une station d'épuration dans l'optique de sa modélisation globale. Master's Thesis, CPE Lyon, Lyon.

Edde, H.J. and Eckenfelder. Jr., W.W. (1967). Theoretical concept of gravity sludge thickening: Scale-up laboratory units to prototype design. *J. Water Pollut. Control Federation*, 40(8), 1486–1498.

Fitch, B. (1966). Current theory and thickener design. *Ind. And Eng. Chem.*, 58, 18–28.

Fitch, B. (1979). Sedimentation of flocculent suspensions: State of the Art. *AIChE*, 25(6), 913.

Hassett, N.J. (1958). Design and operation of continuous thickeners. *Ind. Chem.*, 34, 116.

Kos, P. and Adrian, D.D. (1974). Gravity thickening of water treatment plant sludge. In *94th Annual Conference*. AWWA, Boston.

Merta, H. and Ziolo, J. (1986). Calculation of thickener area and depth based on the data of batch-settling test. *Chem. Eng. Sci.*, 41, 1918–1920.

Nadeau, C. (2016). Modélisation de l'épaississement de la boue rouge par cisaillement. PhD Thesis, Université du Québec, Chicoutimi.

Obata, E. and Watanabe, Y. (1979). Analysis of semi-continuous thickener using a batch settling curve. *H. J. Chem. Eng. Japan*, 12(4), 307–312.

Rudman, M., Simic, K., Paterson, D.A., Strode, P., Brent, A., Šutalo, I.D. (2008). Raking in gravity thickeners. *Int. J. Mineral Process.*, 86(08), 114–130.

Rudman, M., Paterson, D.A., Simic, K. (2010). Efficency of raking in gravity thickeners. *Int. J. Mineral Process.*, 95(10), 30–39.

Suttill, K.R. (1991). The ubiquitous thickener evolves into several complementary designs. *Eng. Mining J.*, February, 20–26.

Talmage, W.P. and Fitch, B. (1955). Determining thickener unit areas. *Ind. and Eng. Chem.*, 47, 38–41.

Tiller, F.M. (1981). Revision of Kynch sedimentation theory. *AIChE*, 27(5), 823–829.

Vesilind, P.A. (1968). Influence of stirring in the thickening of biological sludge. PhD Thesis, University of North Carolina, Chapel Hill.

Water Environment Federation. (1998). *Design of Municipal Wastewater Treatment Plants*, 4th edition. WEF Press, Alexandria.

Waters, A.G. (1989). Theory and application of thickener design. In *Dewatering Technology & Practice Conference*, Brisbane, 81–89.

Wilhelm, J.H. and Naide, Y. (1981). Sizing and operating continuous thickeners. *Mining Eng.*, 33, 1710–1718.

4

Sludge Treatment

In terms of both volume and nuisance, sludge is the main waste product to be treated at a wastewater treatment plant. In addition to organic carbon, the removal of nitrogen and phosphorus results in an increasingly large mass of sludge that must be treated.

Treatment involves four phases:

– reducing the water present in the sludge to decrease its volume;

– stabilization of organic matter (OM) to prevent fermentation and putrefaction problems;

– obtaining a suitable texture to make the sludge manageable and portable;

– recycling.

4.1. Sludge origin

All wastewater treatment processes result in the production of waste. This waste is collected:

– at the pretreatment stage: screening rejects, waste from skimming and sand removal;

– at the primary, secondary and even tertiary settling tanks: the sludge.

Generally, wastewater treatment in a wastewater treatment plant involves four successive stages:

– Pre-treatments: these involve removing coarse elements (screening), sand (sand removal) and grease (grease and oil removal).

– Primary settling: this allows the capture of settleable suspended solids (SS).

– Biological treatment: this reduces the wastewater OM load through the use of microorganisms.

– Sedimentation (or clarification): this allows the separation of the bacterial "floc" from the aqueous phase.

The treated water is then discharged into the natural environment, while the residual sludge is collected and then treated for reuse or disposal. All urban sludge is of the hydrophilic organic type, thanks to its high content of OM or volatile solids (VS) and protein and its good fermentability.

The sludge produced by wastewater treatment plants consists of undegraded OM, mineral matter, microorganisms and water (approximately 99%).

4.2. Sludge quality

Characterizing sludge from urban wastewater treatment allows us to determine its performance (dryness, capture rate, etc.), consumables (reagents, electricity, air, water, etc.) and applicable costs (in volumetric flow rate or dry matter (DM) per surface area or linear meter). For each unit of sludge treatment, the goal is to determine the best compromise between investment costs (related to the load) and operating costs (related to consumables).

A material such as sewage sludge is extremely complex, both in terms of its chemistry and composition and in terms of its physical properties, which notably determine its mechanical and thermal behaviors. Therefore, each sludge is unique, and it is not easy to predict the behavior of a given sludge under a specific treatment with a good confidence interval without the thorough characterization of the sludge.

4.3. Composition of the sludge

Urban sewage sludge is composed primarily of fertilizing elements, notably phosphorus and nitrogen. Table 4.1 shows the nature of the sludge and its concentration. In addition to these elements, it also contains:

– Various heavy metals, the main ones being: cadmium (Cd), chromium (Cr), copper (Cu), mercury (Hg), nickel (Ni), lead (Pb) and zinc (Zn). Analyses show that they are present in trace amounts, but sometimes some of them are found at concentrations toxic to microorganisms.

– Organic micropollutants, such as polycyclic aromatic hydrocarbons (PAHs), polychlorinated biphenyls (PCBs), pesticides, drug residues and industrial by products.

– Pathogenic microorganisms: viruses, bacteria, protozoa, parasitic worms and fungi.

– Therapeutic treatment substances, including hormones and, in particular, contraceptive substances and cancer treatment residues (AMORCE 2012).

Parameters	Primary sludge	Biological sludge	Mixed sludge	Physicochemical sludge
Origin	Primary settling	Secondary clarifier	Primary and secondary treatments	Tertiary treatment
Composition	Solid particles	Organic by-products	Mixture of primary and secondary sludge	Chemical hydroxide flocs Mineral particles
	Gray color Concentration: 20–40 $g.L^{-1}$	Brown flocculated material Concentration: 10–20 $g.L^{-1}$	Concentration: 30–60 $g.L^{-1}$	Concentration: 40–50 $g.L^{-1}$

Table 4.1. *Composition of sludge from a wastewater treatment plant*

4.3.1. *General characteristics*

Parameters	Units
DM, SS, VSS	% mass, $g.L^{-1}$ or % DM
Sludge index (SVI or SSVI)	$ml.g^{-1}$
BOD_5, BOD, TP, TKN	$mg.L^{-1}$
Elemental composition C, H, O, N, S	% mass of VM
Lower heating value (LHV)	$kcal.kg^{-1}$ of VM
pH, temperature, voltage	°C, mV
VFA alkalinity in $CaCO_3$	$mg.L^{-1}$
Heavy metal content	ppm of DM
Fat content	% of DM
Fiber content	% of DM
Germ and parasite levels	$germs.ml^{-1}$

Table 4.2. *Chemical characteristics of the sludge*

Sludge characteristics can be chemical and physical in nature. They guide the manufacturer in choosing the best treatment process and the possibilities for reuse. Table 4.2 lists the variables to be identified.

4.3.1.1. *DM*

This is the main parameter considered when defining the treatment process. It determines the quantity of sludge to be treated, regardless of its concentration level within the treatment process. For purely physical treatment, it remains constant except for the capture rate. It is determined using a 105°C oven or infrared spectroscopy.

4.3.1.2. *SS*

While DM is easy to determine in concentrated phases, the same is not true for clarified phases (overflows, filtrates, centrates, etc.), where the procedure for measuring SS by membrane filtration is more appropriate. Caution is therefore advised when calculating the capture rate, which should preferably be expressed as a function of SS.

4.3.1.3. *Volatile matter (volatile suspended solids (VSS) expressed as a % of DM) or VM (concentration)*

This parameter provides valuable information on the degree of sludge stabilization and its suitability for various treatments (dewatering, incineration). The lower the value, the better the sludge's ability to thicken or dewater, but the lower its exothermicity during incineration.

First, we consider that VM = OM, with VM typically being analyzed in a 550°C furnace. At this temperature, the calcination of carbonates can be disregarded (complete at 900°C).

4.3.1.4. *Sludge index (SI) or sludge volume index (SVI)*

This value depends on the retention time in the biological tank. It is indirectly related to DM and VM. Knowledge of this value is particularly important for thickening. The lower it is, the better the sludge's ability to thicken.

Furthermore, the following relationship has been established linking the limiting concentration of extracted sludge X_L at SVI:

$$X_L = 1968\ SVI^{1.106}$$

4.3.1.5. *Chemical oxygen demand (COD), biological oxygen demand (BOD), total phosphorus or phosphates (TP), total kjeldahl nitrogen (TKN)*

Knowledge of these parameters is secondary in the sludge treatment chain. However, it is important to know the values of the various supernatants, overflows, filtrates, etc., returning to the beginning of the water treatment process.

4.3.1.6. *CHONS*

This value allows us to determine the capacity of sludge to be degraded biologically (anaerobic digestion with biogas production) or thermally (incineration) by calculating the degradation stoichiometry. The CHONS value is generally expressed relative to the VM or even relative to the degradable VSS only if biological stabilization is a consideration.

4.3.1.7. *Lower calorific value (LCV)*

This value is related to the CHONS value. Its importance is paramount in incineration. Generally expressed relative to VM, it should not be confused with the higher calorific value (HCV) measured using a bomb calorimeter.

The LCV is fairly well correlated with the COD (in $g.kg^{-1}$) and VSS (in % DM) such that the following correlations can be proposed, expressing the LCV in kcal/kg relative to the DM:

$$\text{LCV} = 3.81\ \text{COD} + 28.6$$

or:

$$\text{LCV} = 41.3\ \text{VSS} + 887$$

4.3.1.8. *pH, temperature, redox potential (rH) and conductivity*

These parameters have no predictive value in sludge treatment, but they can highlight malfunctions such as sludge aging that degrade the performance of the treatment process (drop in rH).

4.3.1.9. *Volatile fatty acids and $CaCO_3$ alkalinity*

As before, these are mostly just parameters for controlling the proper functioning of a unit operation: digestion.

4.3.1.10. *Heavy metals*

Evolving legislation regarding waste discharge makes knowledge of these materials essential, regardless of the final disposal method. The following metals must be identifiable: Pb, Cr, Cu, Mn, Ni, As, Cd and Hg. Furthermore, it is important to understand their inhibitory properties in the event of biological treatment (digestion, aerobic stabilization or composting).

4.3.1.11. *Fats*

Generally expressed as HEM (hexane-extractable matter), these components are included in VM. This content is taken into account in all biological or combustion operations. However, it can be useful to know them in order to take preventative measures against clogging (piping, sensors, etc.), crusting (digesters) or even oxygen transfer deficiencies (aerobic stabilization, stripping, etc.).

4.3.1.12. *Fiber*

At the submicron level, fibers can decrease the specific resistance of sludge and therefore improve its dewatering. However, they constitute a significant portion of the non-degradable fraction of solid waste that must be taken into account, for example, during digestion.

4.3.1.13. *Germs and parasites*

While coliform bacteria are only an indicator of stabilization, the helminth egg content is paramount in assessing the suitability of agricultural land transfer, particularly in the case of vegetable crops (migration of parasites into the roots of plants consumed raw).

4.3.2. *Specific characteristics of the sludge*

Sludge can take on different states related to its rheology:

– Liquid: any sludge up to the dewatering stage (< 10% approximately).

– Pasty (or plastic): any dewatered, or even partially dried, sludge (approximately 10 to 30–35% DM). This is referred to as shovelable sludge.

– Solid: sludge which, when piled to a height of 1 m, forms an angle of at least 45° with the horizontal. This is then considered stackable sludge.

– Dry: dried, powdery or granular sludge (generally with more than 90% dry solid content).

Dryness alone is not sufficient to characterize sludge. Its appearance and pelletability are important for its handling.

Figure 4.1 shows the typical boundaries of the different states. The consistency of sludge is better the higher its VS content, as this consistency is directly related to density.

Table 4.3 shows the physical measurements.

Purely hydraulic or thermal parameters are relevant at different stages of the treatment. In contrast, mechanical parameters primarily concern the thickened sludge before dewatering.

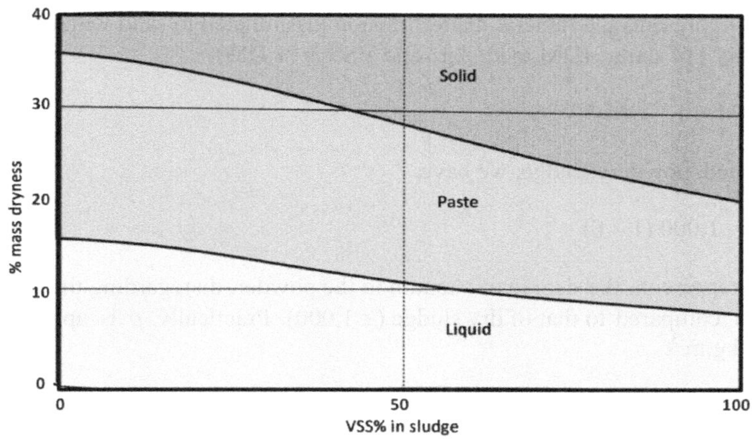

Figure 4.1. *State of consistency of sludge depending on its dryness*

Parameters	Units	Symbol
Hydraulic		
Density	kg.m^{-3}	ρ
Rheology (apparent viscosity)	mPa.s	μ
Thermal		
Thermal conductivity	Kcal/(h.m.°C)	CP
Specific heat	Kcal/(kg.°C)	K
Mechanical		
Capillary suction time	s	TSC
Dryness limit	% mass	SL
Specific resistance	Tm.kg^{-1}	RS
Compressibility index	–	N
Centrifugation index	–	IC

Table 4.3. *Physical measurements that can be assessed for sludge*

4.3.2.1. *Density*

The density allows us to estimate the sludge volume. Without a measurement, the following relationship can be used for liquid or pasty sludge:

$$\rho = 1{,}000\ (1 - MM) + 2{,}700\ MM$$

where:

– ρ is expressed in kg.m^{-3};

– MM represents the mineral matter fraction (assimilated to sand with a density of 2,700 kg.m^{-3}) of sludge (DM in kg. kg^{-1} and VSS in % DM):

$$MM = 1 - DM * VSS$$

For dried, powdery sludge, we have:

$$\rho = 1{,}000\ (1 - \varepsilon)$$

where ε represents the degree of vacuum in the powder, disregarding the density of air (1.29) compared to that of dry sludge (≅ 1,000). Practically, ρ is approximately 300–600 kg.m^{-3}.

4.3.2.2. *Viscosity*

Up to a few tens of g/l in DM, the sludge can be considered a Newtonian fluid, close to water, the apparent viscosity μ being then calculated by an Arrhenius-type relationship as a function of temperature q, the DM content and VSS:

$$\mu = A\ EXP^{\cdot\cdot}[\frac{B}{C + \theta} + D + DM\ (E\text{-}VSS)]$$

A, B, C, D and E are the experimental constants to be determined.

μ therefore decreases with temperature and increases with DM and VSS contents.

When the DM content reaches higher values, the sludge generally adopts a shear-thinning, or even thixotropic behavior, which can be modeled by a Waele-type relationship (de Waele 1929) after destruction of the thixotropic state:

$$\tau = C\ \Upsilon^{n}\ \tau o$$

with:

– τ being the shear stress (Pa);

– τo being the plasticity threshold (Pa);

– γ being the shear rate (s^{-1});

– C being the consistency index ($Pa.s^{-n}$);

– n being the viscosity index.

The apparent viscosity is then:

$$\mu \text{ app} = \tau/\gamma$$

4.3.2.3. *Thermal conductivity (K)*

The thermal conductivity of the dry sludge is approximately 0.4–0.6 kcal/kg. m.°C.

4.3.2.4. *Specific heat (Cp)*

Knowing this value is necessary in biological stabilization or drying. It can be averaged by taking Cp = 1 for the aqueous sludge fraction and Cp = 0.25–0.35 kcal/kg/°C for the DM fraction.

4.3.2.5. *Capillary suction time (CST)*

CST indicates the dewatering capacity of sludge. It is often more relevant to express it as CST/DM. This criterion is not absolute; however, even though it is easily related to the specific strength of a given sludge. Its ease and speed of implementation nevertheless allow for rapid testing of polymer content, for example.

4.3.2.6. *Dryness limit (S_∞) and sludge dryness (S)*

The S_∞ measurement taken simultaneously with the Rs measurement allows for a good confidence interval prediction of the dryness limit at the centrifuge or belt filter outlet.

By measuring S_∞ under 4 bar pressure, the following experimental correlations have been established (correlation coefficient = 0.86), enabling a close approximation of the actual dryness level at the machine outlet S:

$$S = 0.84\, S_\infty + 0.8\%$$

S and S_∞ are given in %.

4.3.2.7. *Specific resistance (Rs)*

This is a reliable criterion for sizing a filter press (FP). Three specific resistances (0.5, 2 and 5 bar) are generally measured in the laboratory, which also allows us to determine the compressibility index and thus optimize the pressure in the filter/filtration time ratio.

4.3.2.8. *Centrifugation index (Ic)*

This is fairly quick to implement. As before, an experimental correlation (correlation coefficient) was established (0.82):

$$S = -2.2\ Ic + 36.2\%$$

Sludge treatment is deemed possible if its Rs value under 0.5 bar is less than 4 Tm/kg. Otherwise, the addition of other reagents is necessary.

The most relevant parameters, depending on the treatment being considered, are summarized in Table 4.4.

Operation	Equipment	Relevant parameters (excluding sludge origin)
Thickening	Static bar screen	DM, SS, VM
	Dissolved air flotation unit	DM, SS, VM, SI
	Gravity dewatering table	DM, SS, VM
	Centrifuge	DM, SS, VM, SI
Stabilization	Digester	DM, VM, CHONS, μ
	Aerobic stabilization	DM, VM, μ
	Lime stabilization	DM
Dewatering	Centrifuge	DM, SS, VM, μ, Sl, Ic
	Belt filter press	DM, SS, VM, Sl, μ, Rs,
	Filter press	DM, SS, VM, Sl, Rs,
Composting	–	DM, VM
Drying	Convective or contact sludge dryer	DM, VM, CP, K
Incineration and flue gas treatment	Sludge incinerator	DM, VM, CHONS, LCV, Cl, F, Heavy metals

Table 4.4. *Relevant parameters to measure depending on treatment*

4.4. Treatments

Upon exiting wastewater treatment plants, sludge contains approximately 95–99% water. Sludge treatment therefore primarily involves reducing its water content and effectively decreasing its pollutant and fermentable load. The objectives of sludge treatment are to obtain:

– a stabilized state since sludge can produce unpleasant smells;

– a sanitized state, measuring the content of germs that can contaminate sludge (enteroviruses, salamanders, helminth eggs);

– a solid texture, for example, generating stable 30° piles of sludge for on-site or end-of-field storage.

The aim is therefore to prepare the sludge for its final stage of recovery or disposal. Five complementary techniques can achieve these objectives: thickening, centrifugation, mechanical dewatering, drying and stabilization-sanitization.

4.4.1. *Thickening*

This step is discussed in detail in Chapter 3 with regard to gravity, lamellar or dissolved air flotation thickening. Simple to implement and energy efficient, it primarily serves to reduce sludge volume and is a preliminary step to subsequent treatments. The dryness level obtained can reach up to 10% DM.

Thickening processes concentrate sludge by removing some of the free water. The finished product thus retains the liquid and free-flowing nature of the feed sludge, so it can still be pumped. Thickening typically increases the DM content to 4%–6%, according to the following estimates: unthickened sludge ~2% DM and thickened sludge ~6% DM.

The processes differ in their equipment, implementation and operation.

4.4.2. *Centrifugation*

Centrifugation is used for both thickening and dewatering sewage sludge, with dewatered sludge having a higher concentration of DM. The centrifugation technologies used for each of these two applications are almost identical.

The main operational differences between the two functions are:

– the rotation speed used;

– the flow rate;

– the nature of the concentrated solid product generated.

Dewatering requires more energy than thickening, because more water must be removed to achieve higher solid concentrations.

This technique involves separating the liquid and solid phases using centrifugal force by adding a polymer to the sludge, resulting in good sludge compaction. Centrifuges have evolved the most in recent years, now allowing similar or even equal performance to FPs. Under the effect of centrifugal force, the SS undergo accelerated settling and are compacted on the wall of the centrifuge. Centrifuges consist of a horizontal cylindro-conical portion rotating at high speed inside which a helical extraction screw rotates at a different speed.

This screw scrapes and removes the solids towards the conical end, while the liquid phase is removed at the other end.

The centrifuge is sized according to the sludge feed rate, its SS concentration and the final conditioning (higher or lower dryness).

Centrifugation performance is significantly improved with the use of a polymer, selected according to the settling test results.

It is necessary to regulate the flow rates and concentrations to ensure consistent centrifuge operation. To achieve this, buffer tanks are installed on the settling tank purges and filter backwash water lines.

This is a fast technology – capable of processing 66–100 m^3 of sludge per hour – and compact, but very energy intensive (150–300 kWh.t DM^{-1}) and very sensitive to sludge quality. The dewatered product, whose DM content can reach 25%, is in the form of a cake: a deformable semi-solid that forms lumps rather than a freely flowing fluid.

Mechanical sludge dewatering processes allow the transformation of liquid sludge (dryness < 10%) into pasty sludge (10% < dryness < 25%) or even solid sludge (30% < dryness). A centrifuge can achieve dryness levels of approximately 20–25%. The addition of a flocculating polymer to further enhance this process (4–6 kg.t DM^{-1}) is necessary. The addition of lime is essential to reach 30% DM content and to dispose of the resulting residue in a landfill or as a calcium-based by-product.

The centrate is returned to the head of the treatment plant via the septic tank. The centrifuges typically rotate at 5,000 rpm.

As with thickening, the most common type of centrifuge used for dewatering applications is the solid bowl centrifuge, generally called a “settling centrifuge”. Its dewatering and solid recovery performance depends on the quality of the feed sludge and the dosing conditions.

The centrifuge can adapt to difficult-to-treat sludge: for oily urban sludge, almost all insoluble matter is separated, dewatering can operate continuously and the process is fully automated and closed, making it more compact than an FP. The main advantage of centrifuges lies in their suitability for small wastewater treatment plants, a sector that FP manufacturers are beginning to invest in. Care must be taken to ensure that the sludge does not contain sand; otherwise, the centrifuge risks rapid deterioration due to abrasion. Therefore, continuous monitoring of the sludge composition is necessary. Furthermore, investment and operating costs (significant consumption of polyelectrolytes and energy) remain quite high. Although its performance has improved considerably, centrifugation only achieves a limited dryness level of between 20 and 25%.

Centrifuges, as they are becoming more energy-efficient, compact and efficient, are the most common dewatering equipment. There are different types of centrifuges, including conventional and high-performance models that produce higher dryness levels but only in the presence of polymers. Performance is linked to increasing the bowl speed, the retention time in the bowl and the screw technology. The main causes of malfunctions originate from the conveyor screw, which must be regularly checked for wear, or from the polymer injection equipment.

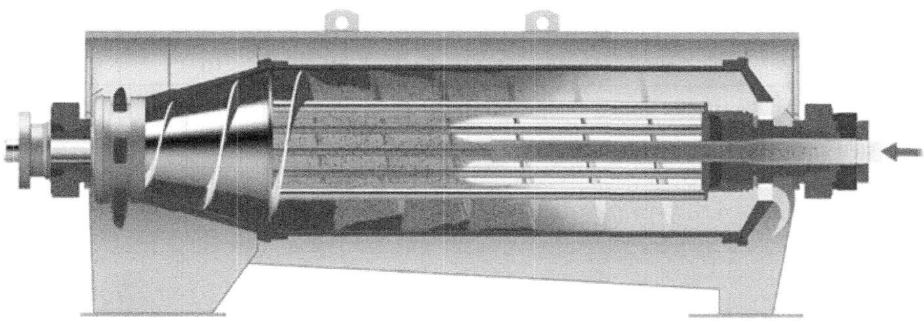

Figure 4.2. *Diagram of a centrifuge*

For correct operation, it is essential to clean the centrifuge after each shutdown and to check all its components. Equipment maintenance requires qualified operators.

The dry solids content obtained is approximately 16–25% for activated sludge. In contrast, for primary sludge, the dry solids content is approximately 50%, and it is heavier, non-flocculated and with settleable solids that facilitate centrifugation. Costs depend on the size of the installation. It is essential to factor in the high costs for energy consumption and the use of consumables.

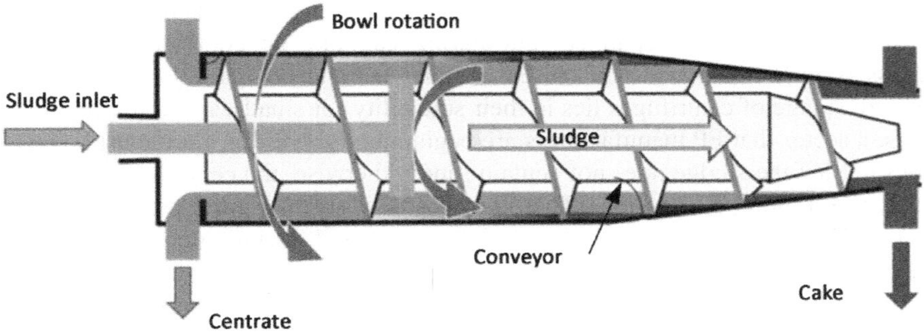

Figure 4.3. *Simplified diagram of operation of a centrifuge (Alfa Laval)*

Origin	Dryness at inlet (%)	Cake dryness (%)	Polymer dose ($g.kg^{-1}$ DM)	SS Capture (%)
Primary sludge	4–8	25–50	2.5–5	≥ 95
Activated sludge	1–2	16–25	7.5–15	≥ 95
Digested primary sludge	2–5	25–40	4–6	≥ 95
Aerobic digested activated sludge	1–3	18–25	10–15	≥ 95
Primary and activated mixed sludge	3–5	25–35	2.5–8	≥ 95
Digested and activated primary mixed sludge	2–4	22–35	7.5–15	≥ 95

Table 4.5. *Drying of dewatered sludge with polymer injection (Andreoli et al. 2007; Metcalf & Eddy 2014)*

Technology	Energy ($kWh.tDM^{-1}$)	Conditioning	Dryness (%)
Gravity thickening	10–20	–	2–3
Flotation	60–100	–	3–5
Centrifugation	150–300	Polymer (5 $kg.t\ DM^{-1}$)	20–25

Table 4.6. *Performance and energy consumption of thickening techniques*

4.4.3. *Mechanical dewatering*

Dewatering processes apply significant mechanical force to achieve greater water removal than is possible through thickening. These processes are as follows:

– pressing the sludge to force the water (or filtrate) through a permeable medium (belt filter, screw press (SP) or FP);

– rapidly rotating the sludge in a cylindrical tank to force the denser solids towards the inner walls of the tank and the water towards its center, the water forming the dilute centrate flow (centrifugation);

– passing the sludge through a narrow, parallel-flow channel with porous walls through which the filtrate passes (rotary press (RP));

– allowing the liquid to evaporate under ambient conditions (sludge drying lagoons) or to evaporate and flow using gravity through a permeable medium on which the sludge rests.

Both thickening and dewatering processes yield a concentrated, consolidated product – retaining most of the solids from the original sludge – and a dilute stream composed primarily of water. For both types of processes, the dilute stream is typically returned to the wastewater treatment plant inlet.

4.4.3.1. *Dewatering technologies*

Dewatering is the second stage of reducing the volume of thickened sludge, stabilized or not, in order to increase the dryness level (on average between 20 and 30% depending on the nature of the sludge).

Dewatering offers several advantages: it facilitates storage and therefore reduces transport costs; it improves stabilization (increased retention time) and facilitates use in agriculture.

There are several mechanical dewatering techniques in the FP category: plate filters, belt filters, membrane filters and SPs.

4.4.3.1.1. Plate FPs

The FP (Figure 4.4) is a dewatering technique that involves mechanically applying high pressure to the sludge. This forces the sludge to release its interstitial water through a filter. A cake of varying dryness then forms along with the retained solids.

An FP dewaters sludge by pressing it between a series of porous plates. The process expels water from the sludge by applying high pressure to the sludge layers held between 20 and 80 rectangular plates. The plates are recessed to allow for sludge filling, and a filter cloth with an effective pore size of less than 0.1 mm is installed on each plate.

The FP is the only mechanical thickening or dewatering technology limited to batch operation (rather than continuous). It is also the only dewatering technology capable of consistently achieving high concentrations of dewatered solids in the final sludge cake between 35 and 45% DM depending on the feed sludge origin and the chemical conditioning applied.

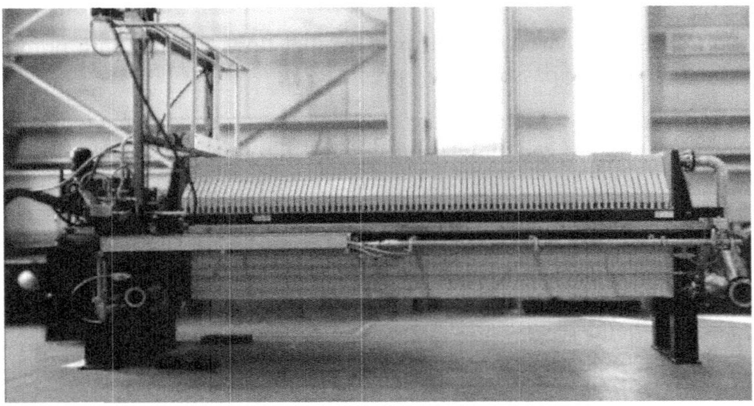

Figure 4.4. *Example of a filter press (source: Cy-Bo)*

The process involves injecting sludge laden with solids under pressure (5–15 bar) and pressing it between two plates. These plates are lined with a filter cloth fixed and suspended vertically between compression plates or compaction membranes. The FP operates by slowly filling the spaces between the plates with sludge and then applying pressures of 7–20 bar for a period of 1–2 hours. Water is expelled from the sludge due to the applied pressure, and the filtrate is returned to the wastewater treatment plant inlet.

When the filter is closed, the plates touch and form chambers designed to receive the limed sludge. This sludge is injected under pressure through the central orifice of the plates and fills the chambers. The liquid portion, the filtrate, passes through the filter cloths and is discharged via a drain. Next, the rubber compaction membranes are inflated with water from a hydroelectric plant or with air. The sludge is compressed in the chamber, further dewatering it until a final compacted cake is formed.

Air is then expelled through the system for 5–15 minutes to remove most of the residual water from the filter cake formed in the plate cavities. As the FP opens tray by tray, the cakes fall successively and are directed to storage silos.

The plates are then separated, and the solid cakes, 25–40 mm thick, can fall out. The complete operational cycle of filling, filtering, emptying and washing the press can take up to five hours.

The next step is to wash the filter cloths: the more "biological" the sludge, the more powerful the washing must be. This washing process also generates significant quantities of water laden with sodium sulfate, which is reintroduced upstream of the treatment process. The resulting sludge dilution must be taken into account to determine the plant capacity. This system generally operates discontinuously according to cycles (filling, filtration, deposition, washing) whose durations depend on the nature of the sludge suspensions, the feed efficiency and the conditioning. The quality of the dewatering is graded by how easily the cake can be removed.

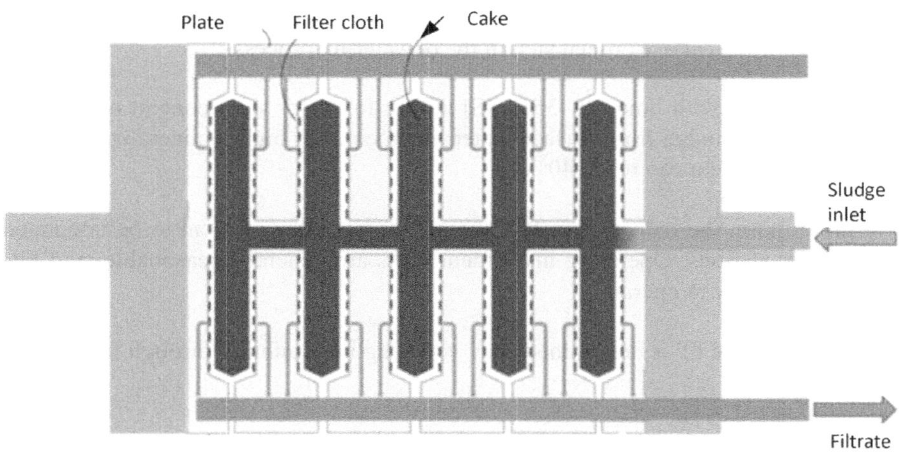

Figure 4.5. *Operating principle of a plate filter press (source: Crystal Lakes)*

Compared to other dewatering techniques, this one requires more labor, particularly due to monitoring the deburring process (the other sequences of the "pressing" cycle can be automated). This constraint can be significantly reduced by using FPs featuring automatic deburring, which offer perfectly satisfactory performances.

Figure 4.6. *Urban sludge dewatered by filter press*

Coagulation/flocculation reagents are used, their main function being to increase particle agglomeration. This facilitates the filtration of these particles. Ferric chloride and lime are the most commonly chosen conditioning agents, but organic flocculants can also be used. The use of specific polymers and fabrics facilitates the application of the cake, a crucial step in the dewatering process.

Conditioning with lime (20–35% SS) allows for a dry solid content of between 30 and 40% for sludge from surface water treatment. For highly mineralized sludge, the dry solid content can reach 40%.

Furthermore, the reduction in the size of FPs has allowed them to be integrated into smaller stations. Operating investment costs are generally reasonable, and FPs are relatively easy to operate.

The size of the FP is determined after verifying its feasibility through filterability tests.

The following parameters should be considered:

– quantity of DM to be processed;

– expected dryness;

– filtration time;

– number of press cycles per day.

The characteristics of FPs are summarized in Table 4.7.

To ensure a consistent supply, buffer tanks on the settling tank purges and filter backwash water should be considered.

	400 kg.h^{-1} filter	600 kg.h^{-1} filter
Filter press		
Filter surface (m²)	35	40
Plate size (cm)	630 x 630	800 x 800
Number of plates	57	52
Number of cakes	56	51
Cake thickness (mm)	25	25
Conveyor belt		
Galvanized steel Power	7 x 0.5 2 kW.ml^{-1}	
Belt speed	1.25 m.s^{-1}	
Belt type	Smooth, abrasion-resistant coated surface	
Pressure pumps		
Piston membrane with pressure regulation	4 m^3.h^{-1}, 5.5 hW	6 m^3.h^{-1}, 5.5 hW
Compressed air membrane	4 m^3.h^{-1}, 6 bars	4 m^3.h^{-1}, 6 bars
Operation	Filling 40 m^3.h^{-1} at 1 bar; pressing 20 m^3.h^{-1} at 6 bars	Filling 40 m^3.h^{-1} at 1 bar; pressing 20 m^3.h^{-1} at 6 bars

Table 4.7. *Characteristics of filter presses*

The main actions are as follows:

– lime preparation;

– cake release control;

– filter cloth cleaning and maintenance.

It is essential to perform complete cleaning once a week to increase the lifespan of hydraulic systems in contact with lime.

The limed sludge is either spread on agricultural land or sent to landfill.

Thickening						**Dewatering**			
Gravitation			**Filtration**		**Accelerated gravitation**	**Filtration**			
Thickened Gravity-fed	Centrifugation	Dissolved air flotation	Thickened Rotating	Thickened Belt	Centrifugation	FP belt	FP screw	FP plate	FP Mb

Table 4.8. *Thickening and dewatering processes (FP: filter press; Mb: membranes)*

Type	Principle	Advantages	Disadvantages	Dryness	Plant type
Belt filter press	Sludge compression between two fabric belts Once the fabric is free of sludge, it must be washed to maintain its porosity This system requires the addition of polymers adapted to the sludge quality	High performance for biological sludge High productivity Continuous operation	Does not process fibrous sludge	25%	Small to medium
Plate filter press	Composed of filtration chambers in which the sludge is compressed by hydraulic cylinders until a compact cake forms The sludge is not fed continuously	Minimal maintenance All types of thickened sludge	Unsuitable for sticky sludge High investment Complicated automation	> 30%	Large
Membrane filter press	Filter press improvement Prevents leakage flow from disrupting or breaking up flocs during the final stage of pressure build-up	Increased dryness compared to a plate filter 40% higher productivity than a plate filter	High upfront cost	30–45%	Limited adoption
Screw press	Spin-drying and compaction phase, using an extrusion screw surrounded by a filter screen. The screens are self-cleaning and specific to each type of sludge	Well-suited for fibrous sludge Continuous operation Low maintenance	Limited dryness	20%	Widely used

Table 4.9. *Technical comparison of different filter presses*

The costs are relatively substantial because they include:

– the filter;

– hydraulic equipment;

– the hopper;

– the conveyor belt;

– the compressor;

– the storage silo.

This technique is quite widespread in large-capacity units despite the discontinuous nature of its operation and its high upfront costs.

4.4.3.1.2. Belt filter press

A belt filter press (BFP) dewaters sludge by pressing it to force water through a permeable medium. The process produces a cake (the dewatered product) with a DM content of 30% or more in the case of primary sludge. This distinguishes it from the low-energy gravity band thickening process, which achieves a maximum DM content of approximately 10%.

A BFP combines drainage and mechanical pressure sequentially to remove water. The thickening process begins with sludge flocculation. Before entering the presses, the sludge is chemically treated to allow the DM to agglomerate into "flocs" and provide initial liquid/solid separation.

Typically, the flocculant is injected upstream of the sludge pump to benefit from progressive mixing caused by the pump screw. This step aims to give the sludge a structure so that it can effectively withstand the pressure of subsequent stages. Therefore, as a general rule, a belt filter cannot function without the addition of polymer.

Next, the flocculated sludge is poured onto the filter belt and spread evenly using a leveling bar: the water released during flocculation flows freely through the mesh of the filter belt. The crucial importance of proper flocculation is readily apparent at this stage. Without prior conditioning, the sludge would simply spread out on the belt and flow down the sides.

The pressing and shearing stages begin at the convergence of the two filter belts. Pressing and shearing occur over a significant length of the belt using several rollers.

This compression action of the two belts in this pressure zone releases more water, allowing for maximum sludge dewatering while maintaining an optimal capture rate. The filtration surface area varies depending on the belt width and the machine model.

The dewatered sludge exiting the final pressing stage is scraped from the filter belts by plastic scrapers and discharged onto a conveyor belt or similar system. Each filter cloth is continuously washed on its return journey to the drainage area to maintain effective filtration.

Most belt filters consist of a low-pressure zone, a medium-pressure zone and a high-pressure zone. The low-pressure zone of the belt filter allows for drainage onto the filter belt. This filtration system operates using pressure as well as a pressing system powered by rollers around which two filter cloths are wound, trapping the sludge. The pressure depends on the size of the belt.

Belt FPs consist of a stainless steel frame, a sludge feed and distribution system, a gravity drainage zone, a pressure (compression) zone, a gear motor drive, a continuous belt washing system, belt tensioning and tracking systems, filtrate collection tanks, integrated piping and electrical controls.

Most belt filters include a thickener that facilitates sludge dewatering.

Some belt FP (BFP) designs are based on a three-belt configuration, in which the gravity drainage belt operates independently from the two pressure belts. This arrangement allows the recirculation rate in the thickening stage to be adjusted independently from the dewatering stage. As with a gravity belt thickener, the belts are continuously cleaned by spray nozzles to prevent clogging of the filter media pores. The wash water is collected and combined with the filtrate.

The overall BFP performance depends on the source and characteristics of the feed sludge, as well as the dosing conditions. As with all other thickening and dewatering technologies, primary sludge – whether raw or anaerobically digested – is more easily filtered than residual activated sludge.

As with all dewatering processes, the concentrated sludge product, known as the cake, is no longer free-flowing and forms solid lumps. It must therefore be handled and transported using a conveyor system or mechanical handling equipment or manually with a shovel.

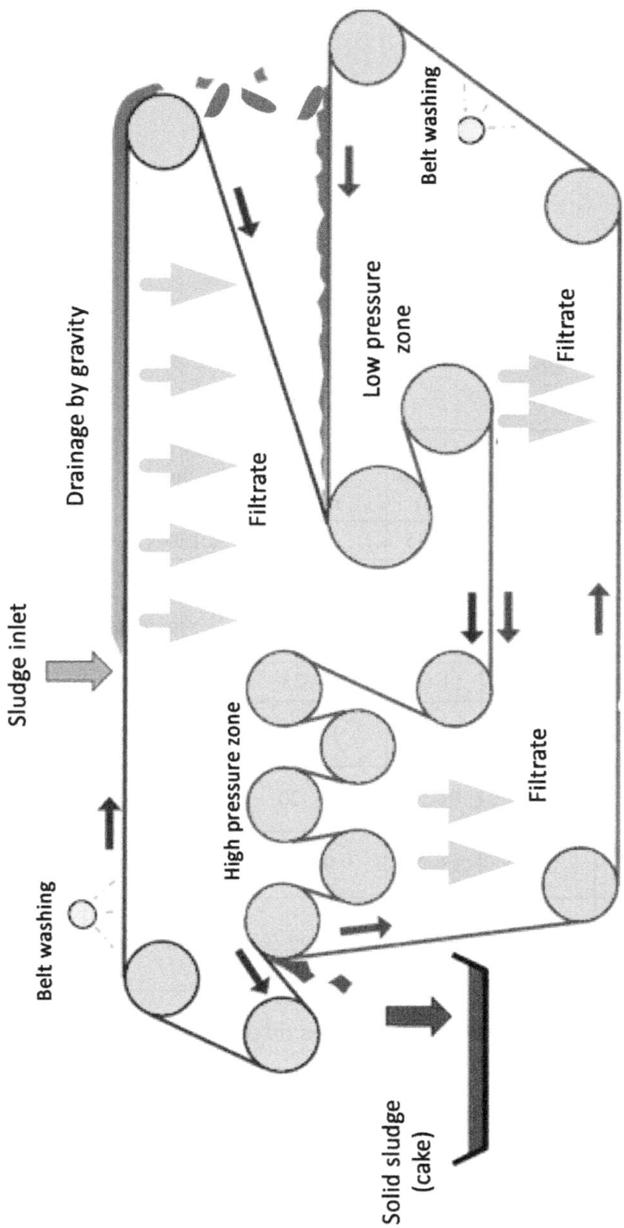

Figure 4.7. *How a belt filter press works (Judd 2023)*

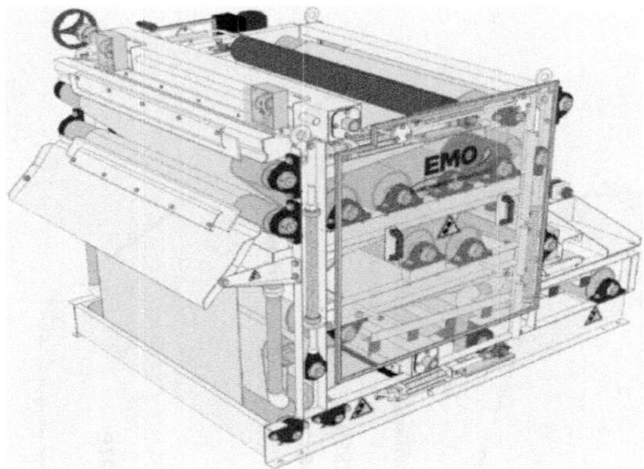

Figure 4.8. *Belt filter press (source: EMO-France)*

Origin	Inlet dryness (%)	Cake dryness (%)	SS/m belt load/ $m^3.h^{-1}$	SS/m belt load / $kg.h^{-1}$	Polymer dose ($g.kg^{-1}$ DM)
Primary sludge	4–8	26–35	14–38	1,130–1,590	1.5–2.5
Activated sludge	1–2	12–20	11–23	180–340	5–10
Sequencing batch reactor (SBR)	1–2	12–19	11–22	250–360	5–7.5
Membrane bioreactor	1–2	11–18	16–25	230–320	5.5–10
Primary digested sludge	2–5	24–35	14–37	680–910	2–5
Digested activated sludge	2–3	13–23	6.6–20	230–410	4–10
Primary and activated sludge	3–5	15–25	9–27	340–820	3–5.5
Digested primary sludge and activated sludge	2–4	1528	9–27	320–540	4×8.5

Table 4.10. *Performance of belt filter presses on urban sludge (Alfa Laval)*

4.4.3.1.3. SP

An SP ensures sludge dewatering by transporting the sludge along a screened cylinder. The process is based on a slow-rotating Archimedes screw (approximately 5 rpm) within a cylindrical screen. The cylinder is typically inclined at about 20° to the horizontal to facilitate water drainage into the sump pit.

The screens are less prone to clogging than the filter media used in other dewatering technologies. Spray cleaning is applied only intermittently for approximately 2–4% of the operating time.

As with all thickening and dewatering processes, the solids content of the product (or cake) depends on the quality of the feed sludge. The solids content of the cake is inversely proportional to the VS content of the feed sludge.

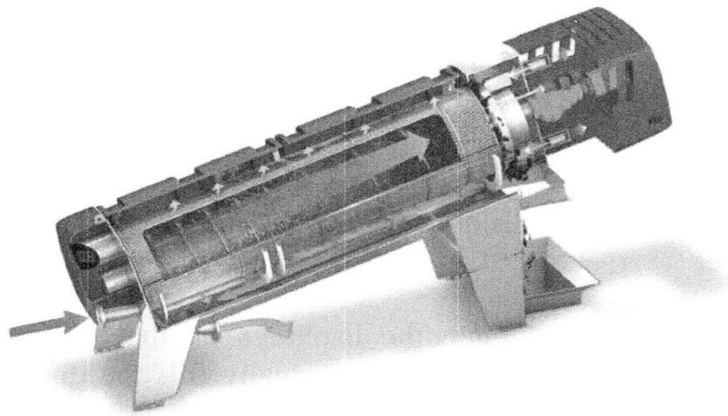

Figure 4.9. *Screw press (source: BTP-corner)*

Origin	Dryness (%)	% capture	Polymer dose (g.kg^{-1} DM)
Primary sludge	30–40	< 90	4–10
Activated sludge	15–22	88–95	8–11
Mixed sludges (primary and activated)	25–35	< 90	5–10
Primary digested sludge	22–28	< 90	10–17
Aerobic digested activated sludge	15–20	88–95	8–17
Digested primary sludge and activated sludge	17–25	< 90	10–17
Activated sludge digested anaerobically	15–25	88–95	8–17

Table 4.11. *Performance of screw presses (Andreoli et al. 2007)*

4.4.3.1.4. Synthetic membrane FP

The sludge is pumped to a sludge concentration tank. A polymer (polyacrylamide type) is mixed with the sludge at a dosage of approximately 0.2%.

The precise formulation and the dosing of the synthetic membrane FP significantly improve the bonds between solid particles, leading to more efficient separation during dewatering.

After thorough mixing, the sludge is pumped into the membrane FP.

The dewatered sludge exhibits good dryness (30–45%).

The filter plate is fitted with a filter cloth and, in the case of a membrane press, an elastic membrane (typically polyurethane (PU) or similar material). The sludge suspension is pumped into the filtration chambers and subjected to pressure generated by a hydraulic closing system and the feed pump. During filtration, the liquid fraction (filtrate) passes through the filter cloth and is collected by the drainage system, while the solids are retained and progressively form a filter cake within the chamber.

Compressed air is used and enters the diaphragm chamber inside the filter plate via an inflation hose. The compressed air gradually fills the diaphragm chamber, causing the diaphragm to expand from the inside.

In some configurations, membrane FPs can perform a secondary squeezing stage to further increase cake dryness and solids recovery. This is achieved by inflating the membrane – typically using water or compressed air – thereby applying additional pressure to the filter cake. At the end of the filtration cycle, the plates are separated and the dewatered cake is discharged.

Regular inspection, cleaning and maintenance of the filter cloths and membrane plates are required to maintain filtration efficiency and extend equipment service life. Membrane plates are commonly manufactured from rubber, polypropylene (PP), PU or composite materials, depending on the chemical and mechanical requirements of the application.

The feed rate is slow and unsuitable for highly concentrated sludge. These FPs are well suited for small treatment plants. The filtration cycle of a membrane FP consists of four main stages: feeding, filtration, membrane pressing and discharge. The duration and efficiency of each stage are affected by factors such as sludge characteristics and operating conditions. Generally, the entire filtration cycle of a membrane FP can range from 30 minutes to 2 hours, depending on the sludge being processed.

– Feeding stage:

  - Duration: usually 10–30 minutes.

  - Influencing factors: sludge concentration, viscosity, feed pressure and speed.

– Filtration stage:

- Duration: usually 10–20 minutes.

- Influencing factors: membrane material and pore size.

– Membrane pressing stage:

- Duration: usually between 10 and 30 minutes.

- Influencing factors: press pressure, sludge properties and membrane performance.

– Cake discharge stage:

- Duration: usually 5–15 minutes.

- Influencing factors: cake adhesion, filter plate design and discharge system efficiency.

The compression pressure in membrane FPs depends on the characteristics of these filters. For medium-pressure membrane FPs, the pressure range varies between 1 and 1.6 MPa (approximately 10–16 bar). For ultra-high-pressure membrane FPs, the pressure range exceeds 2 MPa (over 20 bar). These are used for sludge with very high solids content.

Membrane FPs enable efficient processing of suspensions and drying of solid filter cakes through precise process control and optimized operating methods. By optimizing the filtration cycle and compression pressure, the membrane FP ensures high efficiency and stable operation.

4.4.3.1.5. RP for sludge dewatering

An RP is a dewatering technology that removes water by passing sludge through a channel with porous walls. The RP achieves dewatering using a cylindrical chamber housing two slowly rotating circular screens (< 2 tr/min) assembled one against the other. The sludge enters the channel available between the two screens and follows a circular path around the channel under low applied pressure (0.1–0.5 bar). The combination of applied pressure and shear forces generated by the rotational action, along with the moving solids, facilitates the passage of the filtrate through the two sieves as the sludge moves through the channel. The operation and performance of this technology are similar to those of the SP, which also operates at low speeds and uses a fine screen to retain solid sludge. Unlike the SP and many other dewatering technologies, the RP screens cannot operate with simultaneous filtration and spray cleaning. Furthermore, the filtration area of an RP is practically limited by the unit's diameter (up to 1.2 m). RP installations comprise several parallel units; the technology is modular in nature. It requires polymer dosing.

The RP requires very little energy. Thanks to its fully sealed system, it offers excellent odor control and operates at low noise levels.

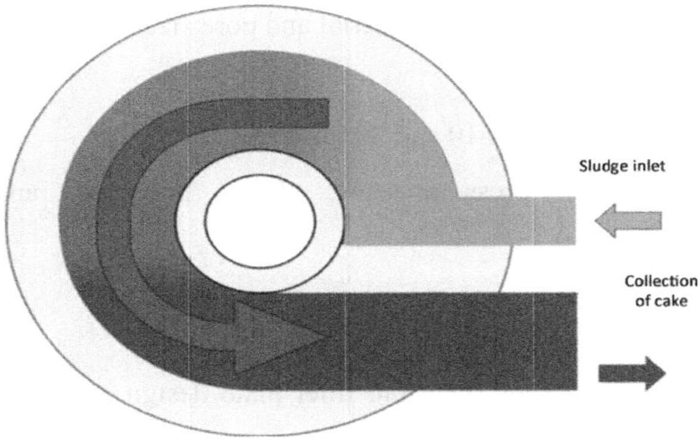

Figure 4.10. *Rotary press (Andreoli et al. 2007; Metcalf & Eddy 2014)*

Table 4.12 presents the performance of RPs with respect to urban sludge. It should be noted that the cake dryness varies between 12% and 45% depending on the type of sludge treated. The SS capture rate is between 90 and 98%.

Origin	Dryness (%)	SS capture (%)	Polymer dose ($g.kg^{-1}$ DM)
Primary sludge	28–45	> 95	2–6
Activated sludge	13–18	90–95	12.5–17.5
Digested primary sludge	22–32	90–95	7.5–10
Digested activated sludge	12–17	85–95	10–17.5
Mixed sludge (primary and activated)	20–32	92–98	7.5–10
Digested primary sludge and activated sludge	18–25	90–95	10–15
Aerobic digested activated sludge	28–45	90–95	8.5–17.5

Table 4.12. *Performance of rotary presses on various types of urban sludge*

4.4.3.1.6. Comparison of three dewatering processes

The dry solid (DM) content of dewatered sludge, or cakes, can vary from 13% to 46% depending on the characteristics of the feed sludge and the dewatering technology used (Table 4.13). Plate FPs are the most efficient and can achieve 33% dry solids content with activated sludge and 46% with primary sludge.

Origin	Filter press	Centrifugation	Belt filter press
Primary sludge	40–46	25–32	31–37
Activated sludge	27–33	14–20	13–20
Digested mixed sludge	29–35	20–25	20–25

Table 4.13. *Dryness (%) observed using different dewatering methods (Andreoli et al. 2007)*

4.4.3.2. *Another mechanized technology: electro-dewatering*

Electro-dewatering technology enhances sludge water removal through non-mechanical means by applying an electric field across the sludge.

Electro-dewatering technology removes water from sludge based on the principle of electro-osmosis. Electro-osmosis is an electrophoretic phenomenon in which liquid moves through a porous solid under the influence of a direct electric field. This movement occurs due to the electrical charge at the solid–liquid interface, which causes water to migrate toward the oppositely charged electrode.

An electro-dewatering device comprises a narrow horizontal channel through which sludge is conveyed. A direct electric field is placed across the channel, with the anode at the top and the cathode and a filter cloth at the bottom. The cathode, on which the filter cloth rests, is porous to allow water to pass through it.

Figure 4.11. *ELODE process from the company Bluewin*

Applying an electric field causes water to be extracted from the sludge under the influence of electrophoretic force. No additional mechanical force is required: the water removal rate depends on the charge on the surface of the particles and the applied voltage.

This method made it possible to obtain a cake containing up to 50% DM, accompanied by a reduction in sludge volume of up to 75% at sludge loading rates of 75 to 150 kg DM.m^{-2}.h^{-1} (per m^2 dewatering zone).

Even though the process appears very efficient, the electrical energy consumption is significant: approximately 300 kWh per ton of water removed. This limits its application to sludge with relatively low water content, that is, conventionally dewatered sludge. As such, it competes with thermal drying as a water removal technology rather than mechanical dewatering technologies whose energy consumption is typically well below 20 kWh per ton of water removed.

4.4.4. *Sludge drying*

Drying is a unit operation in sludge treatment consisting of evaporating free and bound water. This operation follows the dewatering stage. To avoid rapid abrasion of the drying system, the grit chamber in the water treatment stage must be efficient. Its efficiency must be greater than 95% for particles larger than 200 μm.

The advantages offered by drying are essentially:

– a decrease in volume;

– a stabilized and sanitized product;

– a reduction in odor potential under certain conditions;

– easier storage and spreading of the product;

– an increase in the LCV of the sludge.

This solution allows the sludge to solidify by drying it completely; it is then totally dewatered. This solid state increases its volume, facilitating sludge storage and transport.

There are two types of drying:

– thermal drying;

– natural drying.

Regardless of the drying technique used, the resulting sludge has a dryness level up to 95%. Unfortunately, these techniques are very energy-intensive, and solar drying, which is significantly more environmentally friendly, is preferred to reduce the environmental footprint of the sludge sector.

In all cases, these systems are very energy-intensive: electricity consumption is approximately 30–70 kWh per ton of water evaporated and the thermal energy consumption is between 3,000 and 4,000 $kJ.kg^{-1}$ of evaporated water, representing the consumption of 80 L of fuel oil per ton of evaporated water. Therefore, for thermal drying, it is practically essential to use an economical energy source if a favorable final result is desired. It then seems very advantageous to use the biogas produced by the wastewater treatment plant as an energy source for the dryer.

It is strongly recommended to conduct the most comprehensive preliminary study possible of the various stages already undergone by the sludge and its fate in order to assess the relevance of final thermal drying.

4.4.4.1. *Thermal drying*

Thermal drying necessarily follows a mechanical dewatering stage.

It can be convective or contact:

– Convective drying involves the evaporation of sludge by convection, using a heat transfer fluid.

– Contact drying relies on heat exchange by conduction, via a wall heated by a heat transfer fluid.

At the output, the sludge is in the form of powder or granules, with a dryness level up to 90–95%. These two processes are very energy-intensive: they represent an area of opportunity for the possible reduction of the environmental footprint of the sludge sector; for example, by implementing energy recovery loops.

4.4.4.2. *Types of drying*

4.4.4.2.1. Partial drying

Partial drying is intended to raise the dryness of the sludge to an acceptable level for its self-combustion in a sludge incinerator. The resulting product is generally pasty and is neither stabilized nor sanitized.

4.4.4.2.2. Thorough and complete drying

Thorough and complete drying allows for a non-pasty product. Drying to approximately 60% does not produce stabilized sludge nor does it eliminate odors. Complete drying to over 90% produces stabilized sludge that can therefore be stored under cover from rain. One of the often cited advantages of complete or extended drying is the reduction in volume. This volume reduction is indeed very significant for dewatered sludge and sludge dried to 60%. However, the granular appearance of sludge dried to 90% reduces its apparent density, and the volume reduction is not proportional to the amount of water extracted above the paste-like state. One ton of sludge containing 30% DM implies 300 kg of DM.

4.4.4.2.3. Sanitization and stabilization of sludge

Sludge sanitization is recommended when the sludge is intended for spreading. This heat treatment of fresh sludge requires a minimum retention time of 30 minutes at 65°C. Sludge stabilization allows for its prolonged storage. Dried sludge containing sufficient residual water can be recolonized by ambient flora. Fresh sludge is stabilized at DM contents above 90% and therefore only after complete drying.

4.4.4.2.4. Odors

Even complete drying of undigested sludge does not fully eliminate odors. For effective odor control, it is preferable to dry previously digested sludge to approximately 90% dry solids.

Alternatively, if the sludge has not undergone digestion, chemical stabilization – such as liming or nitrite treatment – should be applied prior to advanced or complete drying to minimize odor generation (60 or 90%).

4.4.4.2.5. LCV of sludge

Partial drying of the sludge improves its LCV and therefore makes it self-combustible.

4.4.4.3. *Destination of dried sludge*

4.4.4.3.1. Incineration

The minimum dryness required for sludge to achieve autothermal combustion depends on its volatile matter content and overall composition. For typical municipal sludge, autothermal conditions are generally reached at dry solids contents between 25% and 30%. In cases of co-incineration in a municipal solid waste (MSW) incinerator, dewatered sludge may be fed directly into the furnace using a dedicated injection system. However, introducing sludge with a significantly higher LCV than that of the MSW can disrupt combustion stability and lead to operational issues.

4.4.4.3.2. Agricultural land transfer

In terms of agricultural use, sludge is often sought after for its lime content and fertilizing elements such as nitrogen and phosphorus. In most cases, sludge is dried directly after dewatering, without the addition of lime. The agronomic value of the sludge then lies solely in its OM and fertilizing element content.

Agricultural land transfer often requires stabilized, or even sanitized, sludge that is easy to store between spreading periods. In this case, complete drying is often implemented. However, it is essential to verify that the type of dryer used effectively ensures the sanitization of the sludge (more than 30 minutes at over 65°C) and its stability.

For handling reasons, this dried sludge must be presented in the form of cylindrical or spherical granules. It must be stored under cover to prevent renewed fermentation following humidification by rainwater.

4.4.4.4. *Drying technologies*

Sludge dryers can be classified according to the heat transfer mechanism used.

4.4.4.4.1. Convective dryers

Convective dryers involve a process in which the sludge comes into contact with a heat source. A gaseous fluid is injected directly to enable combustion. This fluid can be air, a mixture of air and combustion fumes or steam. This air, laden with various pollutants, then must be specifically treated. Convective dryers generally cannot produce sludge with a DM content exceeding 70% due to the risk of fire or explosion. Sludge can be dried either on a conveyor belt or in a rotary drum dryer with one to three passes. In this type of dryer, drying is often complete and requires an inlet DM content of over 60% (above the slurry phase). This DM content is achieved by mixing the sludge to be dried with pre-dried sludge.

The sludge to be dried is carried in a stream of hot gas inside a rotating drum. This sludge must be introduced into the drum in a pre-divided and non-sticky form to prevent sticking. Therefore, some of the dried sludge must be recycled. Due to the necessary recirculation of dried sludge, the sludge can only exit the drum after it has completely dried.

Drum dryer

This type of dryer is generally installed for large volumes of sludge that need to be dried (> 3,000 kg evaporated water/h).

The sludge is spread in layers on a continuously moving perforated belt and dried by a flow of hot gas passing through or over the sludge bed.

The sludge deposited on the belt can be sticky, and the drying process may be thorough or complete. With this type of drying, there is no real reduction in volume because the water in the sludge pellets is replaced by pores. Consequently, the resulting granules crumble into dust with the slightest mechanical stress.

Belt dryer

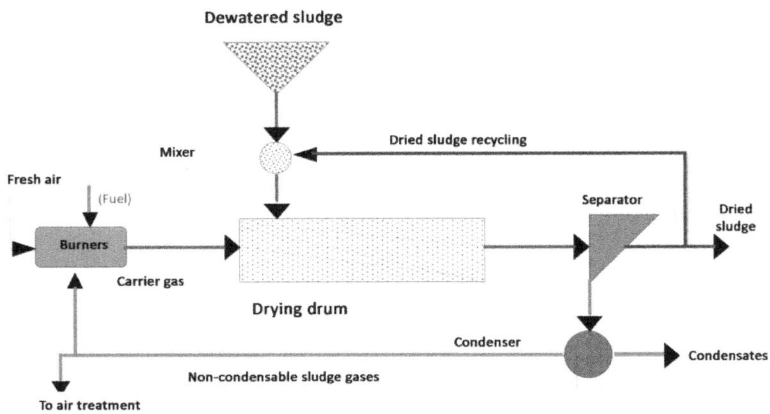

Figure 4.12. *Operating principle of a convective sludge dryer*

Fluidized bed dryer

Fluidized bed dryers allow for complete drying. To obtain good granulation in the bed, the dewatered sludge must be pre-mixed with dried sludge.

The fluidized bed dryer simultaneously removes dust from the dried sludge granules. The dust is carried away with the fluidized air. The granules form in the bed and are extracted, dust-free, from the side of the upper part of the bed. Drying is carried out using heated fluidized air.

4.4.4.4.2. Contact dryers

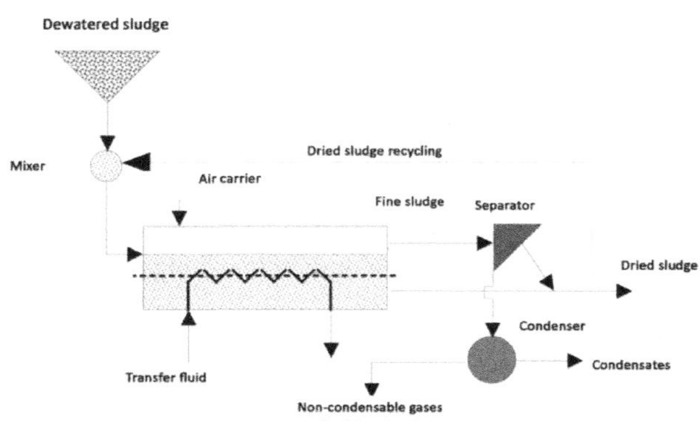

Figure 4.13. *Operating principle of the contact dryer*

Contact drying is a complete drying process. The sludge is heated by a heat transfer fluid circulating in insulated pipes. This is a split-flow drying method. The wall is heated either electrically or by a heat transfer fluid, such as steam or a thermal liquid (pressurized water, oil, etc.).

Contact dryers can safely produce sludge with a dry solids content exceeding 90% and sometimes even 95%. However, all types of dryers require continuous and careful monitoring by qualified personnel due to the risk of damage to the equipment caused by the hardening of sludge with a dry solids content exceeding 60%.

Thin film dryer

The dewatered sludge is fed directly into the dryer. Drying can be partial or complete. Dryness levels exceeding 60% are rarely achieved because the required contact surfaces are significant.

Paddle dryer

The dewatered sludge is introduced directly into the dryer. Drying is carried out to 60% or preferably completely at 90%.

Disc dryer

The sludge must be injected in a non-pasty state: unlike thin-layer and mixer dryers, disc dryers require recirculation.

Fluidized bed dryer

The fluidized bed dryer functions like a convective dryer. The hot gases are heated by coils immersed in the bed.

4.4.4.4.3. Turbo dryer

The turbo dryer is a combination of a convective and a contact dryer. It allows for advanced drying to 60% or 90% and potentially partial drying. The pasty slurry is fluidized into a thin layer (approximately 2 cm) on the cylinder wall. This wall is heated by a heat transfer fluid, and a stream of hot gas flows through the cylinder. A shaft equipped with blades, rotating inside the cylinder, prevents the slurry from sticking to the wall by shredding it and keeping it in a fluidized state along the wall.

4.4.4.4.4. Radiant dryers

In this type of dryer, there is no heat transfer medium. The energy required for drying is supplied directly by electromagnetic or infrared radiation. The dewatered sludge, which must be granulated beforehand, enters the evaporation chamber in the form of pellets conveyed on a belt. These dryers allow thorough drying of the sludge; however, the dried product must no longer be pasty in order to be properly discharged from the belt. During operation, the sludge temperature must be maintained below 90°C to prevent the risk of combustion. This type of dryer offers the advantage of not generating dust during the drying phase and does not require the installation of a dry pellet recirculation system. In addition, these systems are easy to start up and are particularly suitable for low throughput applications (< 500 kg evaporated water. h^{-1}).

Here are the specifications of the Biocon dryer (Veolia) installed at the Randers plant in Denmark:

– Treatment plant capacity: 100,000 p.e.

– Quantity of sludge treated/year: 8,000 tons.

– Quantity of dried sludge: 1,780 tons.

– Percentage of dryness: 90%.

– Unit capacity: 1,750 kg.h^{-1}.

– Evaporated water: 1,350 kg.h^{-1}.

– Installation area: 432 m².

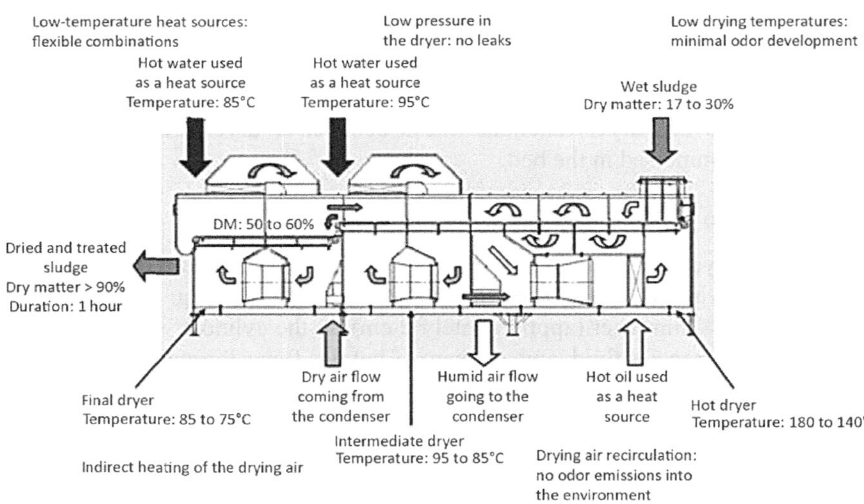

Figure 4.14. *Operating diagram of the Biocon dryer (Veolia)*

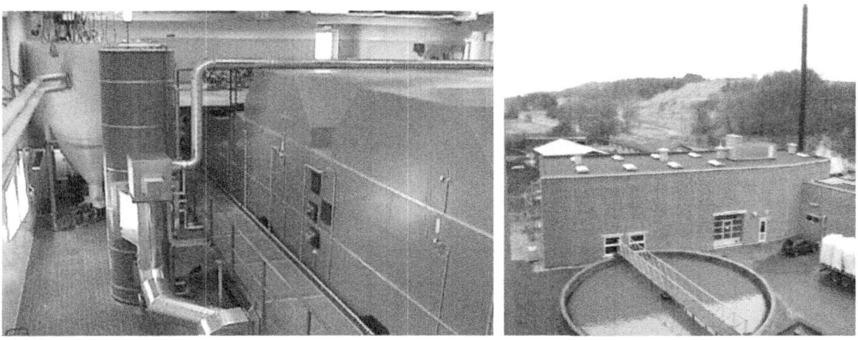

Figure 4.15. *Biocon dryer at the Randers plant (Denmark)*

4.4.4.4.5. Dewatering and drying combinations

In these processes, dewatering and drying are done in the same machine.

Centridry

The centrifuge is positioned in the center of the direct drying drum, while the dewatered sludge is dried by a hot air stream. The sludge cannot be dried beyond 70%.

Start-up and shutdown phases can be very short because the quantity of sludge passing through the dryer is small. This dryer operates continuously.

Drying plate FP

The cake formed during dewatering is dried under vacuum using indirect drying. Vacuum drying ensures that the average temperature of the sludge does not exceed 50°C (the temperature of the sludge varies between 40 at the periphery and 80°C at the center).

The system allows partial, thorough or complete drying of sludge and operates intermittently (batch process). Nevertheless, the maximum temperatures achieved are insufficient to provide thermal sanitization of the final product.

4.4.4.4.6. Characteristics of the products obtained

The properties of dry sludge vary depending on the drying process used and the degree of drying. For example, for 90% dry granules obtained by belt drying, the void ratio is higher than in a drum dryer.

	Convective dryers			Contact dryers				Others			
	Drum	**Belt**	**Fluidized bed**	**Thin layer**	**Paddle**	**Disk**	**Fluidized bed**	**Radiant**	**Turbo dryer**	**Drying filter press**	**Centridry**
Recycling is essential	Yes	No	Yes	Non	No	Yes	Yes	No	No	No	No
Dryer wear	Low	Low	Low	High	High	High	Low	Low	High	Low	High
Thorough drying > 90%	Yes	Yes	Yes	No	Yes	Yes	Yes	Yes	Yes	Yes	No
Thorough drying 60%	No	Yes	Non	Yes	No	No	No	Yes	Yes	Yes	Yes
Partial drying < 60%	No	No	No	Yes	Yes	No	No	No	Yes	Yes	Yes
Dry product	Firm pellets	Brittle, sausage-shaped extrudate	Firm, dust-free pellets	Fine, no granules	Fine, no granules	Firm pellets	Firm, dust-free pellets	Brittle, sausage-shaped extrudate	Firm pellets	Fine, no granules	Fine, large distribution

Table 4.14. *Characteristics of different dryers*

Indicative data:

– Sludge at 35% dry solids: bulk density = 1,020 kg.m^{-3}.

– Sludge at 90% dry solids: bulk density = 600–700 kg.m^{-3}.

– Heat capacity: 0.2–0.4 kcal/kg/°C or 0.84–1.67 kJ/kg/K.

The LCV of sludge is a function of its volatile matter content and the composition of those volatiles. Depending on the nature of the sludge, it varies between 5,000 and 5,600 kcal.kg^{-1} VS or 21,000–23,450 kJ.kg^{-1} VM.

4.4.4.5. *Natural drying*

Drying is a unit operation in sludge treatment that consists of evaporating both free and bound water. Several drying techniques are possible:

– Drying bed.

– Solar drying.

– Sand bed.

– Drying lagoon.

– Bed planted with macrophytes (or reeds).

4.4.4.5.1. Sludge drying beds

Sludge drying beds, along with lagoon technology, represent the simplest dewatering technology. They are generally used directly for stabilized, unthickened solids resulting from prolonged aeration and for digested solids. The principle of the drying bed is to spread liquid sludge over a large area using a bed of gravel and sand. It can be carried out either outdoors or in an enclosed building with mechanical ventilation.

This process involves spreading the sludge to be dewatered over a drainage surface (composed of several layers of gravel and sand of varying grain sizes), through which the interstitial water flows. The sludge is solid upon exiting the drying beds and has a dry solids content of approximately 35–40%. This drying process also offers the advantage of providing a sludge storage solution. It is particularly well suited to wastewater treatment plants serving smaller communities requiring a capacity lower than 5,000 IE.

Drying beds are simple to design and use, as they simply involve spreading the sludge in a thin layer (up to 300 mm) on a porous bed of sand and gravel. Water can then flow through the bed and into an integrated drainage pipe network for collection. Additional water removal is achieved through evaporation from the sludge surface. The degree of dewatering depends on the drying time and, if the bed is uncovered, on the weather conditions. Once sufficiently dewatered (cracked, dark brown surface), which takes 10–15 days, the solids (30–40% dry solids) are mechanically removed. Removal is normally carried out using a front-end loader or specialized mechanical sludge removal equipment. This technique has the disadvantage of being dependent on the climate, relatively long drying times and high labor costs. It is only suitable for large treatment plants due to the surface area required.

Thickened sludge drying beds

The layer of thickened sludge applied is generally between 20 and 30 cm thick. At least two units are required to operate in the filling and drying phases. Polymer conditioning allows for the production of a good quality supernatant within 4 hours.

The operation is broken down into several phases:

– transfer of thickened sludge to the bed;

– sludge settling for 4 hours;

– extraction of the floating material via a sluice gate to a recovery sump;

– the supernatants, still heavily loaded, must be reintroduced into the thickener;

– drainage and initial evaporation for a few days before refilling;

– after n refills, depending on the permissible load, the bed is left to dry, while the second bed is refilled.

Drying beds for unthickened sludge

The applied sludge layer can reach a thickness of 1.5–2 m. At least two units are required to allow operation in alternating filling and drying phases lasting 30–70 d. The operating protocol is as follows:

1) Settling and thickening phase: the muddy water is admitted into the structure where the sludge settles naturally. The supernatant is continuously discharged through wall-mounted slide gates. The muddy water supply stops when the filling time specified is reached.

2) Drainage and drying phase: the water level is lowered very slowly (≈ 10 cm/h) by the drainage system to prevent the sludge from resuspending. Drainage is stopped when the water level reaches approximately 10 cm above the gravel, a level sufficient for proper drying followed by removal with a backhoe loader. Once the drying cycle is complete, the sludge is scraped off. The drains are cleaned and refilled with raw wastewater.

The aim here is to place the sludge on a drainage surface to dewater it. When these drying beds are placed in a greenhouse, the dewatering of the sludge produced by natural evaporation is accelerated by exposure to sunlight; this is known as solar drying.

Other variations exist to avoid dependence on climatic conditions, such as macrophyte drying beds.

The foundation slab of the structure is covered by a 10 cm layer of gravel on which the sludge settles. A DN 200 drain is laid within the bed material along the longitudinal axis of the structure and encased in sand or gravel. Valves must be installed at the upper level to act as an overflow and at the lower level for draining the basin.

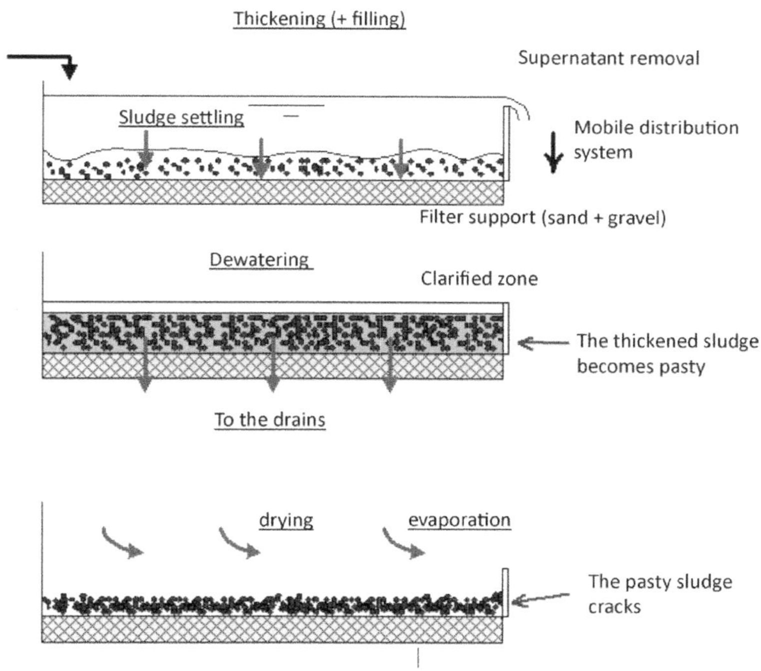

Figure 4.16. *Basic operating principle of drying beds*

The longitudinal dikes have concrete slopes. The transverse dikes have an access ramp for backhoe loaders. The slope of the foundation slab towards the longitudinal axis is 1%.

The "wall" solutions only apply to already thickened sludge. If the sludge layer is thicker (unthickened sludge), the dike solution is mandatory.

4.4.4.5.2. Design sizing

The design sizing depends on the desired sludge quality: at the end of the cycle, the cake must be easy to handle, meaning it must be sufficiently dry and of a reasonable thickness. Generally, short cycles are preferred. For example, British practices (Warden 2003) recommend limiting the final sludge volume to 25 mm under their climatic conditions, which is equivalent to 25 liters of dry cake per m^2, representing a final, and therefore initial, load of approximately 8 kg $DM.m^{-2}.cycle^{-1}$ only for dried sludge up to 30%. About 50 $kgDM.m^{-2}.yr^{-1}$ is the figure often used in Northwest Europe. In Southern Europe, these loads can be up to four times higher.

At the end of the cycle, the sludge must be at least shovelable (corresponding to moderate dewatering) and preferably in lumps (corresponding to advanced dewatering).

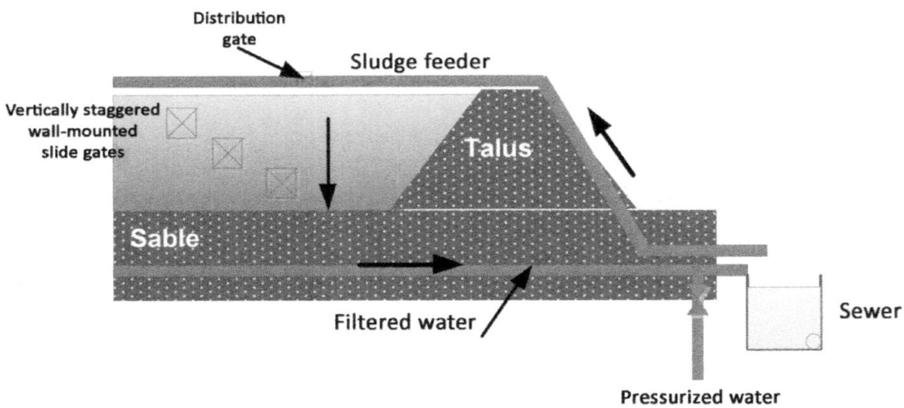

Figure 4.17. *Cross-section of a drying bed*

Whether the sludge is shovelable or in lumps depends on the specific characteristics of the sludge being treated.

4.4.4.5.3. Calculation of loading rate per cycle

The loading rate applied per cycle depends only on the sludge that forms the boundary layer during the percolation phase and not on the climate, which only conditions the final dryness at the end of the drying phase. The combined application of these two constraints defines the design sizing of all the beds. Note that drying climates (hot, dry air + wind speed) ultimately allow for a greater annual applied load than cooler climates because cycles occur more rapidly at equal or greater dryness. A typical cycle is considered to last 10 weeks with a maximum loading rate of 50 kg $DM.m^{-2}$ applied to N – 1 beds, N being between 2 and 3.

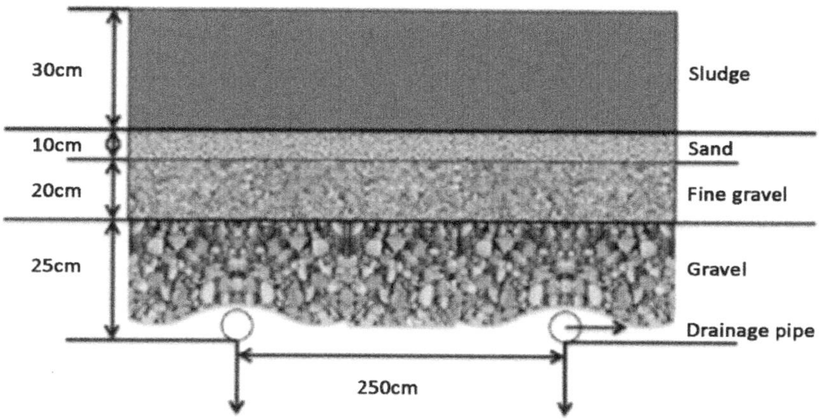

Figure 4.18. *Cross-section of a drying bed with standard dimensions*

Figure 4.19. *Condition of the dried sludge*

The solid loading rate values for drying beds are approximately 100 times lower than those of a gravity thickener.

Origin	Solid loading rate (kg $DM.m^{-2}.yr^{-1}$)
Digested primary sludge	120–150
Primary sludge and digested activated sludge	60–100

Table 4.15. *Applicable load depending on the type of sludge*

4.4.4.5.4. Calculating cycle duration based on climate

In general, the climate determines both the duration of the cycle and the resulting dryness. If the climate is very humid, a cover is recommended. Furthermore, frost significantly enhances dewatering, as freezing promotes the release of bound water.

Climatological data thus defines the expected loads and performance. The water level to be lowered by percolation or overflow is then calculated, taking into account the evaporative capacity of the site. The percolation rate must take into account flow in a porous medium, as removal by overflow depends on the end of the settling phase. If this is not possible, the head is adjusted.

Climate	Cycle duration (during the drying period, in weeks)	Dryness to obtain (%)
Oceanic	8–12	15–25
Continental	8–10	20–30
Mediterranean	4–8	30–60
Arid	2–4	85

Table 4.16. *Cycle duration and dryness depending on climate*

This approach is also applicable to lagoon management, but in this case, there is neither percolation nor overflow. The difficulty in sizing a drying bed is linked to climatic conditions, as data concerning the water balance are often unavailable (Warden 2003).

Below are some simplified calculation formulas, where:

– Xo is the initial dryness;

– Xf is the final dryness;

– DM is the dry mass of sludge.

The wet mass WM to be removed (by settling, percolation or drying) is (mass balance):

$$WM = DM\left(\frac{1}{Xo} - \frac{1}{Xf}\right)$$

where:

– Cm is the mass loading rate of the sludge on the bed (in kg $DM.m^2$);

– X is the concentration of the sludge (in $g.L^{-1} = kg.m^{-3}$).

So, the applied sludge height H is:

$$H = X/Cm$$

Evaporation formula:

– Saturated vapor pressure P° (in mm Hg) at temperature T (in °C):

$$P° = 10^{[7.96681 - 1668.21/(228 + T)]}$$

– Absolute humidity Y° at saturation in kg water vapor/k dry air:

$$Y° = 0.62\ P°/(760 - P°)$$

– Humidity Y for a relative humidity h (between 0 and 100%):

$$Y = h\ Y°$$

Estimating the surface evaporation rate on a body of water: if u is the average air velocity over the water surface ($m.s^{-1}$), then the evaporation rate v $kg.m^{-2}.h^{-1}$ is:

$$v = (25 + 19 * u)\ (Y° - Y)$$

4.4.4.5.5. Impact of rainwater

Depending on the chosen cycle (two beds), each bed is in the filling phase for half the time. During the filling phase, rainfall has no effect, as all of it is diverted to the overflow. At the end of the drying phase, when the first crusts appear, rainwater can be channeled through the crevices to the filter layers and drains. Rainfall will therefore have a significant impact, particularly during the drainage phase, until a cracked cake has formed. This period may represent approximately three-quarters of half a cycle, that is, about four weeks within a 10-week cycle. Consequently, it may be possible to consider operating the system without cover.

EXAMPLE.– Factory in the Middle East.

– Mass loading rate applied = 10,976 kg DM d^{-1}.

– Concentration of incoming sludge: 25 $g.L^{-1}$ (2.5%), that is, 439 $m^3.d^{-1}$.

– Target concentration: 35%.

– Density at 35% DM and 60% VSs = 1.13 kg/dm^3.

– Evaporation = 2,500 mm/yr (2.5 m/yr).

– Maximum height of crust scraping by a mechanized device = 100 mm (= 0.1 m^3/m^2).

– How sludge extraction works: 50 $weeks.yr^{-1}$.

– DM loading rate per cycle: 0.1*350*1.13 = 40 kg $DM.m^{-2}$.

– Water removed by evaporation and drainage of 40 kg DM at 25 $g.L^{-1}$ to 40 kg DS at 400 $g.L^{-1}$ = 40/0.025 – 40/0.40 = 1,500 $l.m^{-2}$, that is, 1.5 m.

– Estimated time for water removal through evaporation and drainage = 6 weeks, of which:

- evaporation = 2.5/50 * 6 = 0.30 m (20%);
- drainage = 1.5 - 0.3 = 1.2 m (80%).

– Total area required = 6 * 7 * 10,976/40 = 11,524 m^2.

– Number of cycles per year = (50/6) = 8.

– Annual DM loading rate = 8 * 40 = 320 kg $DM.m^{-2}.yr^{-1}$.

The main design parameter for drying beds is the initial DM concentration, which determines the total water height to be removed. If this concentration is too low, the drainage effect will be insufficient, and evaporation will not compensate for it. For "easy" sludge (low OM content, high SVI), 80% of the water removed by drainage is considered the maximum achievable value.

Although most drying beds are uncovered and their performance is therefore subject to local weather conditions, water evaporation rates can be increased by sheltering the beds in a glass structure (greenhouse). This is commonly called a "solar drying bed" and is often used in warmer climates.

Although low in terms of energy and operating costs, the large land footprint of drying beds limits their use to applications where land availability is not a constraint.

4.4.4.6. *Solar drying*

Drying beds can be placed in greenhouses to take advantage of natural evaporation, accelerating it with sunlight. This is known as solar drying. Solar drying in greenhouses complements conventional mechanical dewatering of sewage sludge. This process combines the action of solar radiation on the greenhouse with mechanical turning using a self-propelled machine called a “scarifier” or “turner”. Thus, between the greenhouse’s entrance and exit, the air becomes saturated with water vapor, at the expense of the sludge, which dries out. A ventilation system ensures air renewal. Ventilation is provided by openings in the roof, destratifiers (fans located at the top of the greenhouse that create artificial wind) and side vents. This process yields granules with a diameter of approximately one centimeter, which facilitates handling and reduces the volume to be transported and spread. The drying process stabilizes the sludge, resulting in a pleasant odor.

The DM content of the sludge is highly concentrated at approximately 85%. It is rich in fertilizing elements and makes a good nitrogen and phosphate fertilizer.

4.4.4.6.1. Soliamix™ process (Veolia)

Soliamix™ dries and stores sludge in a horticultural-type greenhouse with continuous circulation of dry outside air. When the dewatered sludge enters the greenhouse, it is mixed with the sludge that is still drying and spread into windrows by the Soliamix™ turner. These windrows are periodically turned to renew the exchange surface and aerate them. For an equivalent greenhouse area, the triangular windrow shape offers a larger exchange surface than a sludge bed, resulting in faster drying. It also allows for the storage of a larger quantity of sludge. The reactions that take place during fermentation are exothermic.

Under the influence of solar radiation and thanks to an automated sludge turning system, the water evaporates and is then removed from the greenhouse by a strong airflow. The fermentation stage, facilitated by regular turning of the sludge, accelerates the evaporation process thanks to the energy released by the oxidation reaction of the OM present in the sludge.

Bio-drying and sludge storage processes lead to a temperature increase in the core of the sludge due to biological activity. This temperature rise decreases the residual DM to be removed and promotes partial pathogen inactivation. A significant temperature rise (50–60°C) is observed, which sanitizes the sludge while simultaneously drying it.

This process makes it possible to achieve dryness levels exceeding 85% and to reduce the volume and cost of sludge disposal.

The sludge depth during filling can vary from 10 cm to 50 cm depending on the type of sludge. This drying bed is located in a greenhouse (Figure 4.20), which helps retain heat, resulting in higher air and sludge temperatures, and also accelerates drying by sunlight. To reduce sludge retention time and increase dryness, fans are often installed in the greenhouses. The bed design takes into account the climate, the water balance and the dryness of the incoming sludge.

a) b)

Figure 4.20. *SoliAmix™ a) and Debets Schalke b) turners*

This technique is highly advantageous because it uses a renewable energy source: the sun. Sludge dried using this method, with a high dry solids content exceeding 85%, can be accepted by various energy recovery or disposal channels. Furthermore, the sludge is heat sanitized.

However, this technique also has limitations: its performance depends heavily on the climate, odor production is not zero, and it requires a large installation area.

Figure 4.21. *Solar drying and Soliamix™ (Veolia) process*

4.4.4.7. *Drying lagoons*

A sludge drying lagoon is an option for the dewatering and final disposal of digested sludge in regions where land availability is not a constraint, where the potential environmental impact and risk to human health are limited and where a low-cost solution is needed. This process is similar to a drying bed on large, demanding areas and to the labor-intensive mechanical removal of dewatered solids, but its construction is simpler since no filtrate drainage is required. Lagoons are used in countries with low rainfall, where the expense of chemical reagents for dewatering is not a concern.

These shallow lagoons (approximately 1.8 m deep) operate solely on the principle of evaporation (hence, a watertight bottom to prevent soil pollution). Drying occurs through successive thin layers of sludge, requiring several lagoons to ensure rotation. When approximately 1 m of dried sludge (30–40% DM content) has reached the bottom of the lagoon, it is removed by a backhoe to a post-drying area.

This process is sized according to the site's evaporative capacity (seasonal variation taken into account). It can only be applied to perfectly stabilized sludge to avoid nuisances.

A lagoon comprises a shallow reservoir with a sealed base to prevent the penetration of sludge water into groundwater, aquifers or other environmental water bodies. The lagoon is filled with sludge, and solids are allowed to settle. During the operational drying/dewatering cycle, which lasts several months, water is lost through evaporation, while the solids form a thick layer at the bottom of the lagoon, stabilized by anaerobic biological processes.

Figure 4.22. *Drying lagoons (Judd 2023)*

Typically, a layer of water more than 0.5 m deep is maintained above the digestate sludge layer. This layer allows for the oxidation of odor-producing compounds generated and released by the sludge layer during anaerobic digestion. Once the dewatered sludge reaches the desired solids concentration of 25–30% DM, the dried sludge is removed.

The typical operating cycle of a sludge drying lagoon is generally over one year, depending on the local climate. This is considerably longer than the corresponding period of a drying bed, where most of the water is lost through drainage.

4.4.4.8. *Bed planted with reeds (or macrophytes)*

Another variation of this process involves placing the drying beds under a plant cover (reeds), thus eliminating the need for climatic conditions. This process uses macrophyte beds or reed beds.

The most frequently chosen plants are reeds because of their high leaf area index, which promotes evapotranspiration, and their rapid growth, which requires significant water. These are very advantageous conditions for sludge drying. The sludge to be treated is taken directly from the aeration basin and deposited on the reed bed. This bed consists of a progressively larger particle size (from the surface to the bottom) and is drained.

The drainage of interstitial water from the activated sludge is facilitated by a free annular space that remains permanently present around each reed stem, while SSs are retained at the surface and progressively accumulate.

The proper functioning of the process therefore relies on dense reed growth so that water can be drained throughout the beds. The above-ground parts of the reeds develop from spring to autumn and remain in place during the winter. During this cold period, the roots retain their function.

However, this type of drying requires careful management, both for aerating the spread layers and for supplying sludge. It is often combined with agricultural spreading of some of the sludge deposited on the plant beds. This process requires the use of an additive that allows the sludge to retain its fertilizing properties and reduce odors.

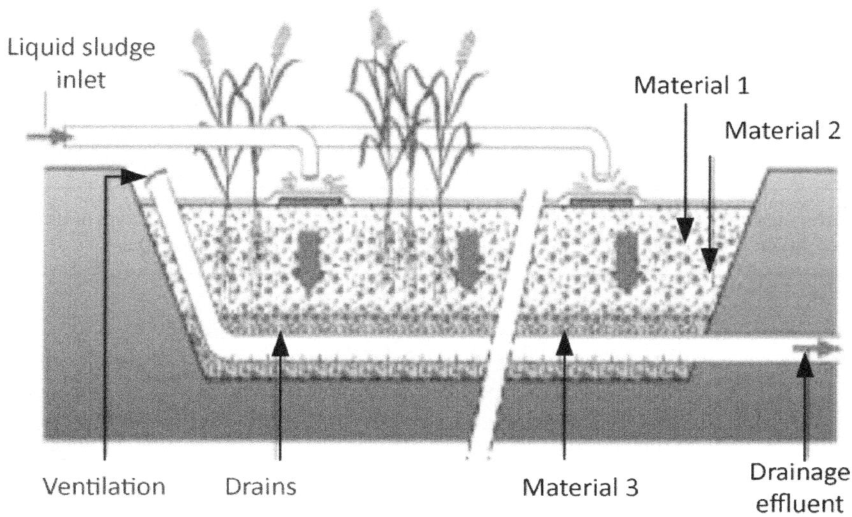

Figure 4.23. *Diagram showing how reed beds work*

These drying beds are designed to stabilize OM and reduce sludge volume to facilitate transport and disposal. They consist primarily of the following elements:

– layers of material with increasing particle size from the surface to the bottom of the filter;

– reeds planted on the surface in a suitable substrate (compost or sand);

– a supply system allowing the direct supply of sludge;

– a freeboard height allowing for long-term sludge storage (1 m maximum);

– a drainage system allowing the collection of drainage water;

– an aeration system allowing the maintenance of aerobic conditions essential for the mineralization of OM.

The parameters considered when sizing the drying beds are the type of sludge to be treated, the quantity of sludge, the applied loads, the number of beds and the supply rate. The sludge is taken directly from the aeration basin and sent to a structure planted with reeds.

	"Economical" solution	Intermediate solution	Technically optimal solution
Top layer	Fine sand + potting soil on 10 cm	Fine sand + potting soil on 10 cm	Fine sand + potting soil on 10 cm
Intermediate layer	Gravel 3/6 or 6/10 on 15/20 cm	Gravel 6/10 on 10 cm	Gravel 6/10 on 10 cm
Geogrid			
Base layer	Gravel 20/40, thickness 0.15, in which the percolate recovery drains are submerged	Concrete floor slabs, 12 cm thick, spaced 3 cm apart	Concrete grating with 2 cm slots, laid on pedestals, 15 cm thick
Comments	Not very favourable for bottom oxygenation	–	Very good ventilation

Table 4.17. *Characteristics of macrophyte-planted beds*

The general sizing conditions are as follows:

– Applied load = 50 kg $DM.m^{-2}.yr^{-1}$.

– Twenty-eight day cycle: 7 days of supply and 21 days of rest with four basins.

The expected dryness performance over a four-week cycle is 10–15% in winter and 15–20% in summer for a Western European climate.

The operating principle is based on the delivery of sludge by truckload. Under the influence of gravity, the water contained in the sludge flows through the materials to the collection drains. SS, meanwhile, are retained on the surface of the sludge mass and form a sludge deposit.

The OM in the deposit thus formed is then progressively mineralized by the action of the microorganisms present. Furthermore, evapotranspiration by plants improves the dewatering of the deposit achieved by simple water drainage (by approximately 10–15%), thus allowing DM content (mass percentage of DM) in excess of 20% to be attained.

Figure 4.24. *Drying bed planted with reeds*

The effectiveness of dewatering and mineralization relies on the biological activity of the bed's constituents: plants and microorganisms.

4.4.4.9. *Geotextile dewatering bags*

For sludge drying, a filter bag drying unit can be used as a replacement for reinforced concrete filter beds.

This system has the following advantages:

– more economical;

– easier operation;

– greater dryness;

– smaller area of land used.

This system uses micro-perforated bag filtration and dewatering systems designed for all types of sludge and requirements below 200 kg/DM/day or 20 m^3/day approximately.

With their very simple modular design, requiring no specific training or special maintenance, it allows for the direct filtering and dewatering of sludge into bags for storage.

Storage allows the dewatering phase to be completed, which involves the reduction of the volume and weight of the sludge. After a few hours of operation, the sludge dryness is approximately 15–30% depending on the percentage of OM to achieve a dryness level of 50–70% of DM after open-air or sheltered storage with natural ventilation.

Figure 4.25. *Teknobag-Draimad® filter bags*

Filtration and dewatering are carried out using special water-repellent fabric bags. The filter media used, insensitive to most chemical compounds, allows air circulation while remaining waterproof for outdoor storage thanks to its specific weave that prevents water from entering. Performance is directly related to the choice of polymer (if required).

For example, the Teknobag-Draimad® 03B-Pack Automatic module in 304 stainless steel is composed of:

– three dewatered sludge storage bags, each with a capacity of 85 liters;

– an open flocculated sludge feed tank (with overflow) above the bags for gravity filling while ensuring in-line sludge/flocculant mixing (maturation and contact time);

– a polymer preparation unit with agitator;

– an adjustable screw pump for sludge feeding;

– an adjustable dosing pump for injecting polymer solution;

– an inline static sludge/polymer mixer;

– a control unit with a digital display and a programmable logic controller (PLC) for pump control, operated by three-electrode level sensors (one/bag) in the bag feed chute, with an adjustable timer, automatic on/off functionality and a standby mode that activates when the bags are full and awaiting replacement.

4.4.5. *Stabilization and sanitization*

Stabilization is a unit operation in sludge treatment aimed at reducing the potential for causing harm. This harm is essentially of two kinds:

– olfactory, due to the putrid fermentations of biodegradable sludge;

– microbiological, due to the presence of potentially dangerous microorganisms.

The second involves pasteurization, disinfection or sanitization of the sludge, obtained primarily through temperature increases. This pasteurization also aims to inactivate viable seeds or pips that are detrimental to agriculture.

There are two main stabilization pathways:

– The biological pathway (anaerobic digestion, aerobic stabilization, composting, etc.) consumes the biodegradable fraction of the sludge and inactivates, at least partially, pathogenic microorganisms through microbial competition. This stabilization therefore results in a reduction in sludge volume.

– The chemical pathway (liming, for example) blocks any microbiological degradation by modifying the physicochemical nature of the sludge (pH, water activity, etc.).

Operation	**Stabilization criteria**
Extended aeration (water treatment)	Sludge age greater than 25 days
Aerobic stabilization of sludge	Reduction of more than 30% in organic matter
Anaerobic digestion	Organic matter reduction: – greater than 55% for primary sludge – greater than 45% for mixed sludge – greater than 25% for excess biological sludge Biogas production: – greater than 500 Nl.kg DM^{-1} inlet for primary sludge – greater than 400 Nl.kg DM^{-1} inlet for mixed sludge – greater than 200 Nl.kg DM^{-1} inlet for excess biological sludge
Liming	pH above 12

Table 4.18. *Stabilization criteria*

It should be noted that sludge stabilization can be advantageously carried out upstream, in the water treatment process, by choosing prolonged aeration of the activated sludge, leading to a significant reduction in degradable volatile matter.

4.4.5.1. *Biological stabilization*

This step minimizes the fresh OM content of the sludge, making it as infertile as possible and thus reducing unpleasant odors.

This treatment is only possible with fresh mixed sludge or secondary sludge.

There are three techniques for stabilizing OM:

– Biological stabilization: aerobic stabilization (in the presence of oxygen) is primarily used to transform sludge into compost for agricultural use.

– Anaerobic digestion: thermophilic anaerobic stabilization, also called "methanation", allows the transformation of sludge into biogas and the recovery of biogas in the form of heat, electricity, fuel or even biofuel.

– Composting: this technique involves chemical stabilization, which neutralizes the biological activity of the sludge to reduce its OM content. To achieve this, the sludge is mixed with lime (quicklime or hydrated lime), the amount of which is calculated based on the sludge's dryness level. This stabilization technique can only be carried out after the sludge has been dewatered. This process is known as sludge sanitization.

4.4.5.1.1. Anaerobic digestion

Digestible OM, whether dissolved or particulate, undergoes the following biochemical reactions:

– hydrolysis of macromolecules into simpler compounds by specific enzymes;

– production of acidic compounds by an acidogenic flora from these elementary molecules;

– gasification, primarily in the form of CO_2 and CH_4, by a methanogenic flora.

The first, rapid acidogenic phase delivers the metabolites necessary for the second, slower methanogenic phase, which is the rate-limiting step of the bioprocess. This first acidic phase must under no circumstances dominate (increasing volatile fatty acids (VFAs) and decreasing pH) because it would then inhibit the second phase.

Sludge digestion is a bioreaction that takes place in a stirred tank. Stirring is achieved either by recirculating the gaseous head into the liquid phase via a compressor or by a mechanical agitator.

Furthermore, a buffer capacity is generally required before dewatering because the digester is not designed to store sludge.

To comply with the minimum pipe diameter ($\geq$ 100 mm) and flow velocity (≥ 0.6 m.s^{-1}) required for proper hydraulic performance, thickened sludge is typically fed to the digester intermittently rather than continuously. The feeding frequency can reach 12–24 feeding cycles per day.

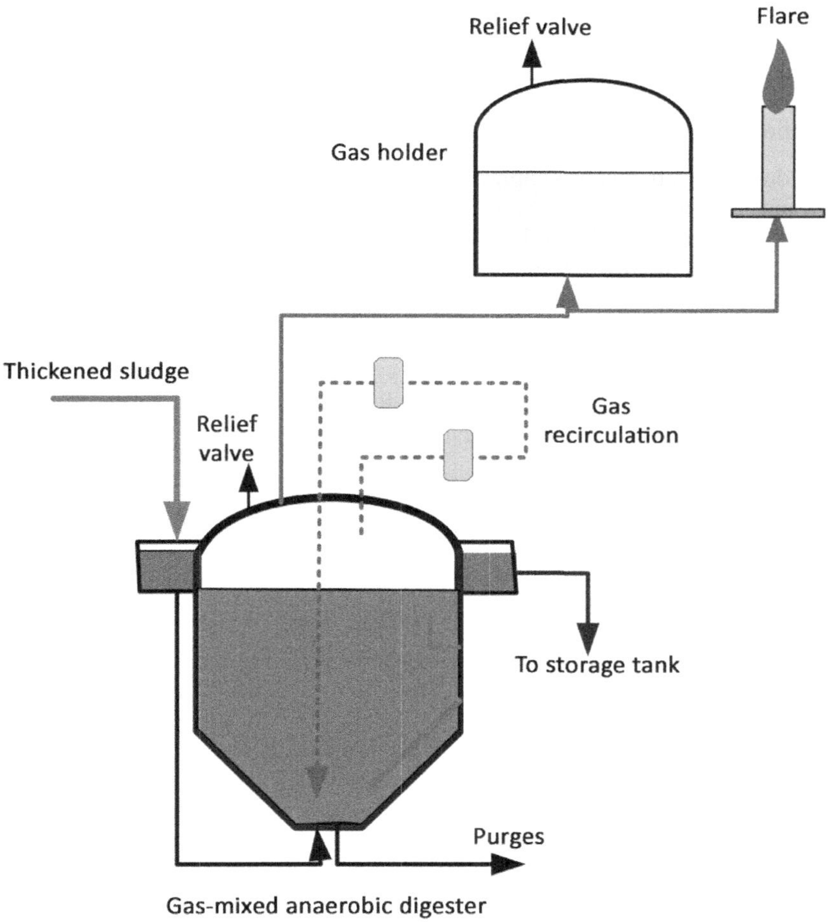

Figure 4.26. *Schematic diagram of a mesophilic gas-mixed digester*

Storage selection and design

Closed storage

This kind of unit has the advantage of being considered, by extension, a "digester–storage unit," since a small fraction of the biogas produced (approximately 5–10%, observed after 15 days of primary digestion and 7 days of secondary digestion) is still recovered at this stage. The unit is not heated. In the case of gas mixing, intermittent agitation can be achieved by diverting the main gas recirculation circuit from the digester. Furthermore, level control over a significant height requires a protective internal lining identical to that used in the digester. The gas holder maintains pressure in the tank during level variations.

Atmospheric storage

The design is obviously simpler, especially for outdoor installations, but this storage unit can produce odors and, if covered, will require appropriate ventilation with H_2S and CH_4 control. Furthermore, there is a risk of surface crusting due to drying. Therefore, intermittent agitation is necessary; its dual function is to break up the surface crust and homogenize the withdrawn liquid.

Performance

These performance levels can be modeled according to the type of sludge. Table 4.19 shows some performance levels observed at numerous stations.

	Primary sludge	**Mixed sludge**
Reduction in OM (or COD) (%)	50–60	45–50
Volume of biogas produced (Nm^3/kg OM destroyed)	0.8–1.1	0.8–1.0

Table 4.19. *Digestion performance according to sludge type*

For mesophilic digestion, reducing the retention time, down to 15 days during peak season, does not significantly affect the sludge removal efficiency (approximately 90% of the maximum observed reduction) but can produce poorly digested sludge with poor dewatering properties and lingering odors, particularly if mixing is inadequate, leading to short circuits and sand deposits that further reduce the actual retention time.

However, an actual residence time of 15 days is considered sufficient to classify the sludge as Class B under US EPA 503 and can therefore be used as a sizing value (European sizing ≥ 21 days at 35°C ± 3°C during peak months, American sizing (EPA class B part 503) ≥ 15 days at 35°C ± 3°C during peak months).

	Mesophile 35°C (30–38°C)	Thermophile 55°C (50–58°C)
Mass load	1.5–3 kg $VM.m^{-3}.d^{-1}$	2–5 kg $VM.m^{-3}.d^{-1}$
Peak hydraulic retention time (calculated based on the digester usable volume)	≥ 21 days	≥ 12 days
Maximum daily load/annual average	1.5	1.3

Table 4.20. *Main operating parameters*

For thermophilic digestion, it is also possible to reduce the retention time to as low as 10 days during peak season. Below this, the system becomes very sensitive to any load variation.

Quality of the sludge obtained

Theoretically "stabilized", the sludge has a composition favorable to agronomic use, as shown in the example in Table 4.21.

	Fresh sludge	Digested sludge
Soluble in ether (fats)	24.5%	10%
Water-soluble (sugars, amino acids and organic compounds)	10%	7%
Soluble in alcohol (resin, waxes, alkaloids)	2.5%	2%
Cellulose and hemicellulose	7%	2.5%
Lignin	6%	10.5%
Proteins	27%	24%
Ash	23%	44%

Table 4.21. *Example of the evolution of the composition of a sludge*

Quality of the biogas produced

Table 4.22 summarizes the chemical composition of the biogas obtained by mesophilic digestion of urban sludge.

Gas	Value	Variation
CH_4	65%	55–70%
CO_2	30%	25–40%
N_2	3%	2–7%
H_2	1.5%	1–5%
H_2S	0.05% (500 ppm)	50–5,000 ppm

Table 4.22. *Biogas quality*

The physical characteristics are as follows:

– LCV for DM = 23,500 $kJ.Nm^{-3}$ = 20,600 $kJ.kg^{-1}$ = 6.5 $kWh.Nm^{-3}$;

– density = 1.13 $kg.Nm^{-3}$;

– specific heat = 1.84 $kJ.kg^{-1}.K$;

– kinematic viscosity = 12.2 10^{-6} Pa.s;

– adiabatic exponent: Cp/Cv = 1.267.

Sanitization

Germ destruction is greater in thermophilic mode, where the temperature is 55°C. We observe a 10_6–10_7 decline in coliforms and a 10_2–10_3 decline for helminth eggs.

	Mesophile 35°C	Thermophile 55°C
Total coliforms	100–1,000	10^6–10^7
Helminth eggs	1–10	100–1,000

Table 4.23. *Germ reduction comparison*

During mesophilic digestion, parasites are therefore likely to contaminate certain crops, particularly vegetable crops.

Process limitations

The limitations of the recommended procedure are:

– the need for a sufficiently concentrated and sufficiently organic raw sludge (> 25 g $VM.L^{-1}$);

– the absence of inhibitors in the raw sludge (careful identification of industrial dischargers connected to the sewer network is essential);

– the significant variation in input load over time (volumetric or mass);

– the return of ammoniacal nitrogen and phosphorus to the head of the plant must be taken into account in the design of the water treatment line.

Fundamentally, it is essential to bear in mind the reductive nature of the biochemical reaction, which alone cannot eliminate H_2S and NH_3.

Sizing of anaerobic digesters

The sizing of an anaerobic digester includes the following elements:

– the tank area (digester and storage tank, gasometer);

– the stirring system (gas or mechanical);

– sludge and gas circuits;

– the heating system (heat exchangers and boiler).

Mass balance

The essential information needed for sizing is as follows:

– the origin and composition of the sludge: type of sludge, DM and VM content, etc.;

– the average daily flow rate Qe (peak month) and daily peak flow rate, if applicable;

– the presence of possible inhibitors;

– the length of stay (usually provided by the project manager) over a peak month or year.

This information allows us to predict expected performance, such as the reduction of volatile matter and biogas production in Nm^3 per kg of volatile matter removed.

Digester shape and stirring type

The shape of the digester (either "tall" or "flat") is defined by the homothetic ratio H/D (height/diameter), as well as its configuration (flat-bottom cylindrical tank or egg-shaped tank with a conical bottom). These shapes depend on site constraints (whether the digester is above or below ground, the water table level, the available surface area, etc.).

The type of stirring chosen is related to the digester shape and size. From an economic point of view, the choice would be:

– mechanical stirring for small (V < 2,000 m^3) and medium-sized (V < 6,000 m^3) digesters, which are tall and ovoid in shape;

– gas agitation for medium- and large-sized digesters, which are rather flat in shape.

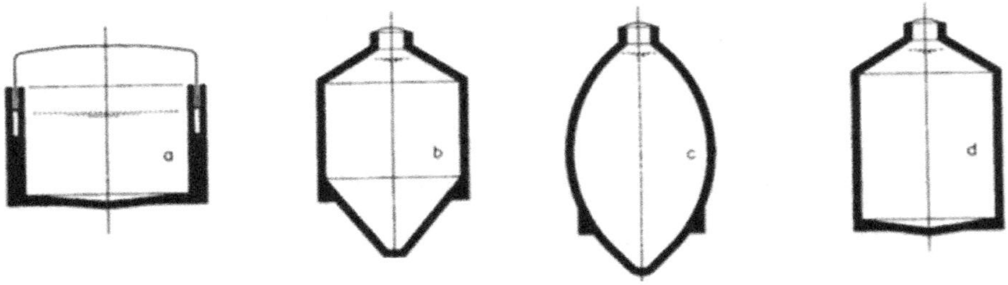

Figure 4.27. *Different types of digesters*

Four configurations are shown in Figure 4.27. The "British" configuration, when mixing is done by gas injection, is unfavorable in terms of mixing efficiency due to its relatively flat geometry, which promotes the accumulation of sand at the bottom and floating materials at the surface. Nevertheless, this design is still widely used because of its simplicity, often incorporating mechanical mixing systems and a less flattened geometry to improve performance.

The "classic" configuration, although providing good hydraulic homogeneity and facilitating sand removal, is relatively expensive due to the foundation costs associated with its conical lower section and the complexity of installing equipment on a steeply sloped structure.

The "egg-shaped" configuration, similar to the previous one, is deemed optimal for mixing: the energy expended is about half that required for the British shape.

The European configuration is used for very large structures (> 10,000 m^3, the world's largest reaching 16,000–17,000 m^3).

Mixing

Homogenization of the sludge in the digester is an important parameter. The choice between gas mixing or mechanical mixing depends on the shape of the digester and the sludge concentration. Table 4.24 presents the advantages and disadvantages of both systems.

	Advantages	Disadvantages
Gas mixing	Mixing in flat-shaped digesters (H/D = 0.5–1)	No sand resuspension Complicated gas circuit Increased risk of foaming
Mechanical mixing	Easy resuspension of sand Suitable for both conical and flat-bottom tanks Low power consumption	Moving parts inside the tank Tangling of fibers Need to stop and empty the digester in case of a mechanical problem Increased risk of corrosion Risk of vortex

Table 4.24. *Advantages and disadvantages of mixing types*

Gas mixing

The Grec/A recirculating gas flow rate is between 1 and 2 $Nm^3/m^2/h$, with Grec/A = 1 for H < 10 m and Grec/A = 2 for H > 20 m and where A is the surface (m^2), Grec is the gas recycled and V is the volume (m^3).

When compared to the usable volume V, the Grec/V ratio is between 0.06 and 0.2 $Nm^3.h^{-1}$ per m^3.

The power dissipated in the tank Pd is then calculated using the following formula:

$$Pd = Grec\ 101325\ Ln\ (1 + \frac{H\ injection}{10.33})$$

Pd is calculated in W if Grec is in $Nm^3.s^{-1}$ and H injection is in m.

Mechanical mixing

Slow mechanical mixing involves one or more large axial-flow propellers mounted on a vertical shaft located in the center of the digester. The following values are used:

– surface velocity (= pumping or primary flow rate/base area) ≅ 0.025 $m.s^{-1}$;

– electrical power consumed = 1–2 $W.m^{-3}$;

– propeller diameter = 0.2 digester diameter;

– peripheral speed < 4 $m.s^{-1}$;

– intermittent operation.

Semi-rapid mixing is ensured by one or more retractable submerged stirrers that can be removed from the tank (digester or storage tank) for maintenance:

– Electrical power consumed: 10–15 Watt.m^{-3}.

– Pumping flow rate ($m^3.h^{-1}$): tank volume x (2.5–3).

– Peripheral speed: approximately 10 $m.s^{-1}$.

– Discontinuous use: 10–15 min every 1–2 hours, with a maximum of 4–6 $h.d^{-1}$.

Rapid mixing in a draft tube is feasible for H/D ratios ≥ 1. The tube is positioned in the center of the structure, along its entire height, to ensure optimal mixing.

Circulation with an external recirculation pump is possible but rarely used due to its very high energy consumption (EPA 1987; Bode and Klauwer 1999).

4.4.5.1.2. Thermophilic aerobic stabilization

Thermophilic aerobic stabilization (TAS) serves either a stabilization function or a predigestion and sanitization function upstream of an anaerobic digester. In all cases, it is located downstream of the thickening stage.

The stabilization function is an alternative to mesophilic anaerobic digestion. In terms of performance (reduction of VSs, agronomic value, etc.), and especially operating costs (electricity), it is less advantageous than digestion. However, due to the short retention times, investment costs are lower, which can make this solution attractive for small treatment plants. In this case, the sludge can be spread directly, without dewatering.

Depending on the predigestion process, this can incur significant additional costs. However, it is advantageous when the client requires sludge sanitization/ pasteurization. Furthermore, predigestion can reduce retention time or improve the actual digestion process and can therefore also be considered in certain rehabilitation projects.

Operating principle

Stabilization of OM through biological oxidation

This kind of stabilization involves biological oxidation carried out by injecting air into a mixed tank. The principle is no different from that used for activated sludge, except for the concentrations and temperature. Above 40°C, the flora involved is no longer the same; nitrifying bacteria cannot operate under thermophilic conditions, which alters the oxygen balance. The following stoichiometries can thus be written according to the temperature conditions.

Under mesophilic or psychrophilic conditions (< 40°C), nitrification, leading to acidification, can occur if the retention time is sufficient:

$$C_5H_7O_2N + 7\ O_2 \rightarrow 5\ CO_2 + 3\ H_2O + NO_{3-} + H^+\ [+\Delta H]$$

Under thermophilic conditions (between 40–45 and 65°C), we have the following stoichiometry:

$$C_5H7O_2N + 5\ O_2 \rightarrow 5\ CO_2 + 2\ H_2O + NH_3\ [+\Delta H]$$

Approximately 1.4–1.5 kg of oxygen is theoretically required per kg of material degraded. Oxidation is incomplete since the nitrogen remains in a reduced form. Under these conditions, the enthalpy of the reaction ΔH can be between 4,000 and 5,000 kcal/kg VS degraded or approximately 3,000–3,200 kcal/kg of O_2 transferred.

Strong ventilation is necessary, and the temperature rise (45–55°C) occurs naturally due to the exothermic nature of the oxidation reactions of OM. The OM content of the sludge is easily reduced to 40–50%.

Batch pasteurization

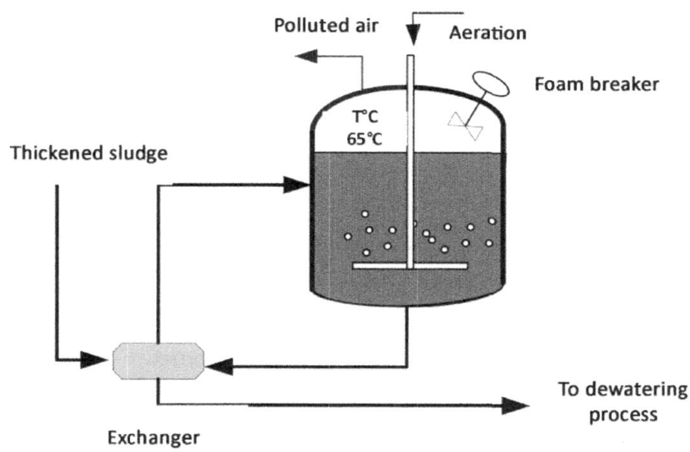

Figure 4.28. *Diagram of how a thermophilicaerobic stabilization system works*

The sludge sanitization/pasteurization function is obtained by ensuring that the sludge mass to be treated is treated for at least 30 minutes at 65°C (or 20 minutes at 70°C). The growth rate of thermophilic flora is optimal at 65°C, then zero above 70°C. Feeding and withdrawal are carried out intermittently via a sludge/sludge heat exchanger (Figure 4.28).

Typical performance and operating parameters

The reduction of volatile matter, a key parameter in sizing, depends on the activity of the aerobic flora. This is classically linked, among other things, to the temperature T in °C and the retention time t in days in the reactor.

An empirical approach, but one that correlates fairly well with reality, consists of calculating the product [T t] and estimating the percentage reduction of volatile matter using the following formulas:

– for (T t) < 230°C.d: % reduction VM = 0.15 (Tt);

– for (T t) > 230°C.d: % reduction VM = 32 + (Tt)/92.

Application	**Aerobic stabilization**			**Predigestion**
Retention time	2 days	4 days	6 days	16 hours
Temperature	60°C	60°C	55°C	65°C
Reduction in VM	16–25%	34–38%	36–40%	6–7%
Reduction in DM	11–14%	22–24%	23–25%	4–5%
Reduction in COD	20–30%	35–40%	38–45%	6–8%

Table 4.25. *Performance of thermophilic aerobic stabilization*

Pathogen elimination

Table 4.26 gives some data for the two main applications.

Types of microorganisms	**Stabilization 6 days at 55°C**		
	Inlet germs/ml	**Outlet germs/ml**	**Reduction**
Enterobacteria	$6.4\ 10^6$	440	$10^{4.2}$
Fecal streptococci	$0.4\ 10^6$	230	$10^{3.4}$

Table 4.26. *Example of pathogen elimination by thermophilic aerobic stabilization (EPA 1979)*

Enterobacteria and fecal streptococci are generally eliminated effectively. Helminth eggs are almost completely eliminated.

4.4.5.2. *Composting*

Compost is defined as an organic amendment comprising carbon compounds of plant origin. The composting of sludge or other organic waste of similar dryness (grass cuttings, for example) occurs solely through aerobic processes and involves a thermophilic phase that destroys pathogens and evaporates a significant amount of water. It is therefore a process that ensures both the stabilization and sanitization of the sludge, as well as its drying. Composting can thus be presented as an alternative to thermal drying.

Any type of sludge can be used, particularly sludge from extended aeration. Conversely, composting digested sludge is not very beneficial, as its biodegradability is lower and its nitrogen content is reduced. Similarly, liming sludge before composting is not recommended, at least at the levels used for stabilization (> 30% CaO/DM). This process leads to a nitrogen deficiency through the release of ammonia and can potentially cause a pH increase that may be inhibiting.

The dryness of the dewatered sludge should preferably be between 15 and 30%. Below this, the moisture content is such that an excessive amount of co-product is required. Above this, the sludge may be difficult to handle or mix (forming non-homogeneous pellets).

4.4.5.2.1. Implementation

Regardless of the technology chosen, sludge composting involves:

– a mixing phase: sludge paste mixed with a co-product in a proportion such that the biological reaction can start;

– a rapid aerobic degradation phase via ventilation;

– a slow maturation phase;

– a phase to refine the finished product.

The aerobic degradation phase is the core of the process, as it is during this rapid aeration phase that evaporation, drying and sanitization occur. Oxygen transfer can take place in the following two ways:

– Ventilated, unturned channels, which are relatively simple to implement, but the need to cover and then treat odors can make them expensive. Typical aeration time: three to four weeks.

– Ventilated channels mechanically turned by a backhoe wheel in a confined hangar. Typical ventilation time: two to three weeks.

Figure 4.29. *Sludge composting*

4.4.5.2.2. Principle

The basic principle consists of aerating a mixture composed of fresh sludge, a bulking agent and possibly recycled compost, followed by a composting phase lasting several weeks. The bulking agent may itself partially originate from internal recycling through trommel screening. A sludge storage area upstream is generally required. Indeed, composting is a batch process requiring manual labor for the various mixing and conveying operations.

Figure 4.30 outlines four possible composting methods depending on whether the focus is on investment, byproduct consumption or compost production.

The structuring co-product has three roles:

– expansion (increased porosity) allowing the natural or forced passage of air;

– water absorption from sludge;

– input of assimilable carbon to improve the overall biodegradability.

In other cases, the mixture may be deficient in nitrogen (limed sludge that has released NH_3), and it is advisable to add an N source in the form of fertilizer (e.g. ammonium phosphate type).

A C/N ratio of 25–35 is generally considered a good value for starting fermentation.

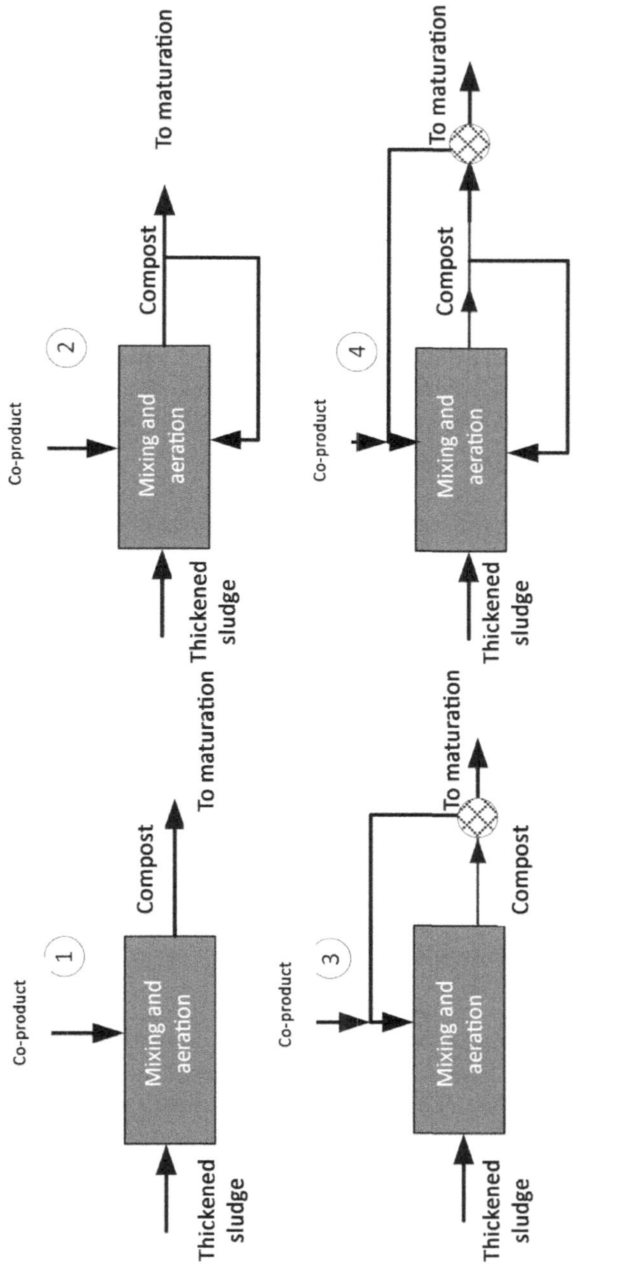

Figure 4.30. *Possibilities for compost production*

The structuring co-product may consist of sawdust, shredded bark or wood chips, green waste or shredded woody pruning residues (leaves and grass clippings should be avoided due to their high moisture content) or any other material – generally lignocellulosic – providing the required structural and bulking functions

Recycled compost also contributes to these three functions, albeit to a lesser extent. It allows for faster reseeding of the compost pile. The mixture then undergoes two to four weeks of aeration. This aeration period can be broken down into three phases, which must be managed according to the moisture content of the mixture being processed:

1) mesophilic phase (< 40°C, typically 37°C): approximately three to four days;

2) thermophilic phase (> 40°C, usually between 55 and 65°C): approximately four to eight days. The temperatures reached allow for sanitization, significant degradation of the solid waste products and substantial evaporation;

3) cooling phase: approximately 7–15 days.

As the biomass sterilizes itself, the temperature decreases (Figure 4.31).

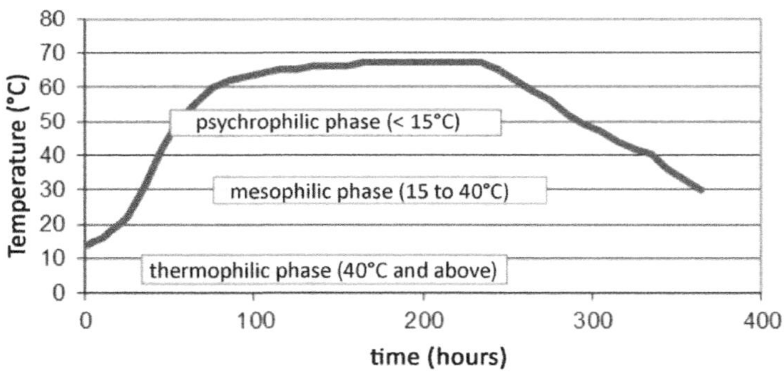

Figure 4.31. *Typical temperature profile*

The pH regulates itself within the optimal neutral range for the ecosystem (6–7.5). Aeration aims to provide the oxygen necessary for the biological oxidation reactions of OM and to maintain an optimal temperature by removing the heat produced during these reactions. The process is managed by measuring the temperature, and even the oxygen content of the piles, and adjusting the daily volume of air injected accordingly. After two months of maturation, intended to reduce residual BOD, there are no more unpleasant odors, the compost structure is light, and the product is perfectly stabilized. Bioavailability to plants can be assessed by cress seed tests, soluble nitrogen (rapidly measured by residual ammonia) and sulfides.

4.4.5.2.3. Performance and operating conditions

The following conditions are recommended for obtaining a stabilized and sanitized product:

– apparent density: 0.40–0.60;

– dryness: 45–65%;

– VM reduction in sludge obtained: 35–55%;

– C/N < 20;

– angle of slope > 45°.

For germ elimination in 21 days:

– total coliforms (germs. mL^{-1}): $10^8 \rightarrow 0$–1;

– salmonella (germs. mL^{-1}): $10^4 \rightarrow 0$–1.

4.4.5.2.4. Advantages and disadvantages of composting

The advantages are as follows:

– The resulting product is dried, stabilized and sanitized (by temperature, nutrient consumption and antibiotic production by fungal flora), thus meeting the most current requirements for sludge treatment.

– Compost is perceived as a rural product because the compositing process is agricultural not industrial. Therefore, any negative impacts are essentially limited to the noise from the agricultural machinery used.

– Versatility and adaptability: a well-sized unit is capable of treating sludge and by products from different sources.

– Energy consumption is very low.

As for the disadvantages:

– It is necessary to provide a co-product, which results in a significant increase in proportional costs and is subject to supply.

– Odor problems can arise with certain types of sludge (mainly primary or mixed).

– It is a biological process: kinetics and performance depend on the variability of the substrates used (sludge and co-product) and the season.

– It is an extensive process: investment costs are high, even for the high-end range (drying, for example, is cheaper in terms of investment costs).

– Although a high degree of drying and mass reduction is achieved, the volume of material to be removed remains significant due to its low bulk density and the addition of a co-product, which increases the final volume.

4.4.5.2.5. Aerated windrow composting

Given the constraints related to handling, the following can be expected:

– a minimum storage capacity of one week;

– a structuring co-product depth of 3–4 m, that is, 1.5–2 m^3/m^2.

The sludge can be stored:

– piled up in a storage area, with a height of approximately 70 cm, that is, 0.35 m^3/m^2;

– in a silo.

When compost is prepared using a by product, a green waste shredder is used. A material size of 50 mm is generally targeted, but the uniformity control is poor, so sizes of up to 150 mm or more can end up in the shredded material.

A minimum windrow volume of 50 m^3 is recommended. Below this, heat loss becomes too significant. The following dimensions are suggested for aeration windrows:

– Leveling height of the compost: 2 m minimum or 2.5–3 m if the handling equipment allows it.

– Unit width of 4–7 m considering an average angle of 40°.

– Windrow length: 15–60 m. Below this, 50 m^3 will not always be achieved; beyond it, there is the problem of even air distribution or the need for multiple fans.

– Maturation: Generally, the retention time should be longer than six weeks. The compost can be stacked up to 3 m high. Considering the leveling, plan a ratio of 2–3 $m^3.m^{-2}$.

– Storage of the finished product: the compost can be stacked between 3.5 and 4.5 m high by the loader. The volume/surface ratio is approximately 3–4 $m^3.m^{-2}$.

4.4.6. *Comparison between the various processes*

Table 4.27 summarizes the operating conditions of the various systems. The criteria for choosing a technology are linked to treatment objectives, sludge recovery, the site, the environment, cost constraints, labor, etc.

	Centrifugation	**Belt filter**	**Plate filter**	**Drying beds**	**Lagoon**	**Reed beds**
Dryness	Good	Average	Very good	Very good	Good	Good
Sludge at inlet	Thick	Thick	Thick	Thick	Slightly thickened or raw +/–	Raw
Polymer	Yes	Yes	Yes	No	No	No
Odors	Yes	Yes	Yes	+/– low	+/– low	+/– low
Noise	High	High	High	No	No	No
Labor	High	High	High	Cyclic	Cyclic	Cyclic
Compactness	High	High	High	No	No	No
Investment	High	High	High	Land	Land	Land
Consumables	Polymer	Polymer	Polymer	No	No	No
Reliability	High	High	High	Very high	Very high	Very high
Weather-related uncertainty	None	None	None	Unfavorable	Unfavorable	Unfavorable

Table 4.27. *Comparative criteria between thickening and dewatering systems*

4.4.7. *Chemical stabilization*

Anaerobic digestion and aerobic stabilization occur in the concentrated liquid phase, that is, after the sludge has thickened. Composting and liming are carried out in the solid or paste phase, with dewatered sludge. Chemical stabilization consists of blocking the biological activity of the sludge by mixing in quicklime, CaO, or slaked lime, $Ca(OH_2)$. Lime quantities are calculated based on the initial and final dry solids content of the sludge, at a proportion of 10–50% of the sludge's dry solids content, which raises the pH above 12. Typically, this stabilization occurs after sludge dewatering. The sludge is then considered sanitized. This practice is generally used in wastewater treatment plants serving more than 15,000 population equivalents (PEs). Sometimes, chemical stabilization is achieved by adding nitrites with an acidic pH.

Like composting and methanization, liming stabilizes and sanitizes sludge. This process takes place after dewatering and involves pouring quicklime (CaO) or slaked lime ($Ca(OH)_2$) on the sludge, initiating an exothermic chemical reaction. It also increases the sludge pH (above 12). Liming is a completely manageable operation with a modest investment cost. However, it is less effective than other techniques in eliminating spore-forming bacteria and helminth eggs.

This chemical process very effectively reduces fecal contamination and the growth of fecal pathogenic microorganisms in the sludge. It also eliminates odors (stabilization), further increases the sludge's dryness (25–35%) and improves its structure (making it more granular) to facilitate transport and storage.

This is the final stage of sludge treatment. The limed sludge can then be used in agriculture or sent for storage but rarely incinerated.

4.4.7.1. *Stabilization with nitrite salts*

Nitrite salt treatment is well-suited, both technically and economically, to small wastewater treatment plants. It provides effective stabilization through the oxidation of odor-causing compounds, inhibits fermentation activity and sanitizes the system (eliminating most fecal bacteria). The process only takes about 2 hours.

This process also increases the dryness of the sludge (by 2–5%) as well as its fertilizing quality, which is particularly advantageous for agricultural use. It could be an alternative to liming. Nitrite stabilization is carried out in the liquid phase, with thickened sludge, without increasing the dry mass to be removed.

4.4.7.1.1. Saphir™ (Veolia) process

The Saphyr™ process is a chemical sludge stabilization process that allows for easy storage while minimizing unpleasant odors. It can also be considered a sanitization process and a technique for improving dewatering by two to five points, depending on the type of sludge and the equipment used. This process is similar to lime stabilization of sludge, but unlike the latter, the treatment is carried out in the liquid phase, preferably on thickened sludge and without the need to increase the final dry mass to be removed.

Furthermore, the Saphyr™ process does not introduce any toxins into the sludge and therefore does not hinder agricultural transfer, which remains the preferred destination.

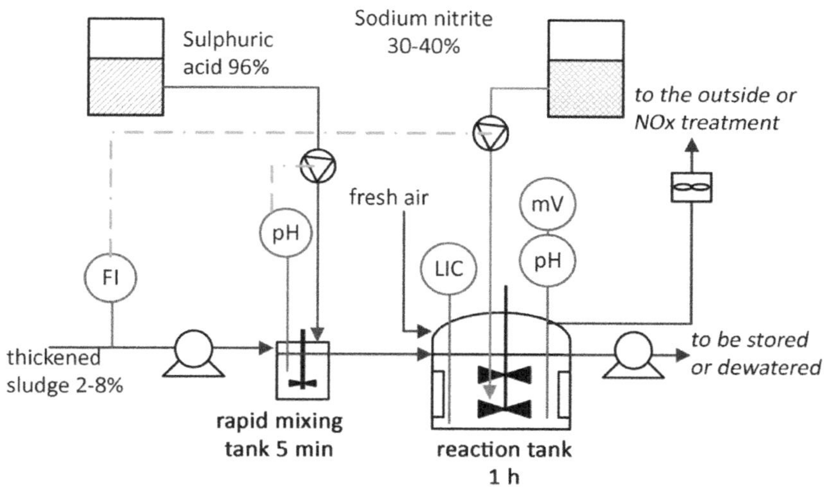

Figure 4.32. *Schematic diagram of a nitrite stabilization unit*

The thickened sludge is transported to an acidic medium maintained at a pH between 2 and 4, depending on the option chosen. It is then subjected to the action of nitrite ions NO_{2-} for 1 hour. A powerful reagent is potentiated in an acidic environment. This results in the destruction of odorous reduced compounds (H_2S, mercaptans, etc.) and also the inactivation of the germs responsible for the degradation of OM. The result is sludge stabilization, allowing it to be stored under cover without emitting pungent odors.

Therefore, pH regulation is necessary in the reactor. Furthermore, a redox probe is required to verify proper operating conditions (eH > 200 mV) and adjust the nitrite flow rate.

Table 4.28 presents two options for nitrite-based chemical stabilization.

	Saphyr™-S (stabilization)	**Saphyr™-SD (stabilization/disinfection)**
Parameters	Stabilization/sanitization Three to six months in storage	Stabilization/sanitization for six to nine months during storage, significant reduction of germs
Target pH in the reactor	2.5–3	2–2.5
Target redox	200–350 mV	> 350 mV
H_2SO_4 qty	100–200 l.t^{-1} DM	250–400 l.t^{-1} DM
$NaNO_2$ qty as rqd for redox	3–4 kg N.t^{-1} DM	4–6 kg N.t^{-1} DM

Table 4.28. *Implementation parameters for Saphyr™ process*

4.4.7.1.2. Partial sanitization of the sludge

At pH 3, after three months of storage, the following reductions in flora can be observed, as shown in Table 4.29.

Germs per 1 g wet sludge	**Initial concentration**	**Saphyr™-S (pH 2.7–3)**	**Saphyr™-SD (pH 2–2.3)**
E. coli	10^5–10^7	5–6 log	5–6 log
Salmonella	10^2–10^5	5–6 log	5–6 log
Streptococci	10^5–10^6	3–6 log	3–6 log
Clostridium spores	10^5–10^7	1 log	4–6 log
Helminth eggs	100	50–70%	50–70%

Table 4.29. *Reduction in bacteria and parasites after treatment*

The main advantage of the low-pH SD variant is therefore the reduction of clostridia and, of course, a longer storage life.

The process allows for an increased dryness of two to six points, provided that an acid-resistant polymer is used.

The volume and dimensions of the tanks are calculated based on the following retention times:

– hydraulic retention time of the rapid mixing tank = 5 min;

– hydraulic retention time of the reaction tank = 60 min.

Acid and nitrite levels depend on the chosen treatment option (S or SD), the type of sludge (primary/mixed/biological) and the sludge's DM concentration. The two reactive agents, sulfuric acid and sodium nitrite (which forms nitrogen peroxide), necessitate the selection of resistant materials. The presence of chloride ions in an oxidizing environment causes pitting on stainless steel.

A 40% nitrite solution is supplied in a 1 m^3 intermediate bulk container. A preparation from a powder solution will be calculated based on 300 $g.L^{-1}$ (saturation threshold: 750 $g.L^{-1}$) in order to limit the temperature drop of this endothermic reaction. The injection of $NaNO_2$ is set to a flow rate proportional to the inlet flow rate to reach the redox setpoint defined in the treatment.

4.5. Sludge reuse

Sewage sludge is prepared for reuse in several stages:

– stabilization;

– drying and drainage to reduce volume;

– dewatering to remove all traces of water.

The sludge can then be reused in agriculture or incinerated. Proven technologies such as anaerobic digestion, thermal hydrolysis, co-digestion and thermal drying allow the sludge to be converted into a valuable energy source. The various methods to recycle sewage sludge include spreading, composting, methanization, incineration, landfilling and alternative processes.

4.5.1. *Energy recovery*

After initial dewatering, the sludge is incinerated in specialized furnaces or mixed with other waste such as household waste and treated in non-hazardous waste thermal treatment facilities.

4.5.1.1. *Thermal treatment*

Dewatered sludge is a difficult waste to incinerate due to its compact texture and high moisture content (55–70%), with a low calorific value. In this context, traditional grate furnaces are poorly suited. Therefore, other combustion processes that promote mixing and oxidation of the material have been developed.

It is advantageous to pre-dewater the sludge (partially) in order to increase its calorific value and reduce the volume to be treated. Generally, due to auto-combustion thresholds, the DM content of sludge treated in a specific furnace is between 25 and 45%.

One of the best-known processes is the Pyrofluid™ furnace, which has a fluidized sand bed (Figure 4.33).

Pyrofluid™ is an advanced, safe and sustainable thermal treatment solution designed for the efficient disposal of dewatered municipal and industrial sewage sludge. The process relies on a fluidized bed incinerator that operates 24/7 at approximately 900°C to oxidize the OM contained in sewage sludge. It produces harmless mineral residues while enabling energy recovery and ash recycling.

Pyrofluid™ can process and ensure the complete combustion of large quantities of sludge – from 200 kilograms to 5 tons of DM per hour (sludge with a dry solid content between 25 and 35%) – from plants of varying capacities, in just a few seconds.

Fluidized bed technology ensures complete combustion with little excess combustion air, and the thermal energy produced offers the advantage of being suitable for many uses, such as building heating or the supply of electrical energy via steam production.

Depending on their characteristics, the ashes produced by Pyrofluid™ can be recovered and used as a replacement for valuable materials such as cement or other natural materials such as kaolin (used in cement plants, precast concrete) or simply disposed of in landfill.

Once dewatered, the sewage sludge is pumped into the fluidized bed furnace at several points. The furnace consists of a layer of hot sand fluidized by combustion air. The sludge is sprayed directly onto the hot sand. Incineration takes place primarily in the sand layer at a temperature of approximately 750°C and continues in the freeboard area, where gas expansion and post-combustion occur and where the temperature rises to 850 or even 880°C. After cooling, the flue gases are treated and purified by an electrostatic precipitator or a baghouse filter, which removes more than 99% of the ash.

Sodium bicarbonate and activated carbon are used to remove acidic compounds and heavy metals. Finally, the purified fumes are vented through the chimney.

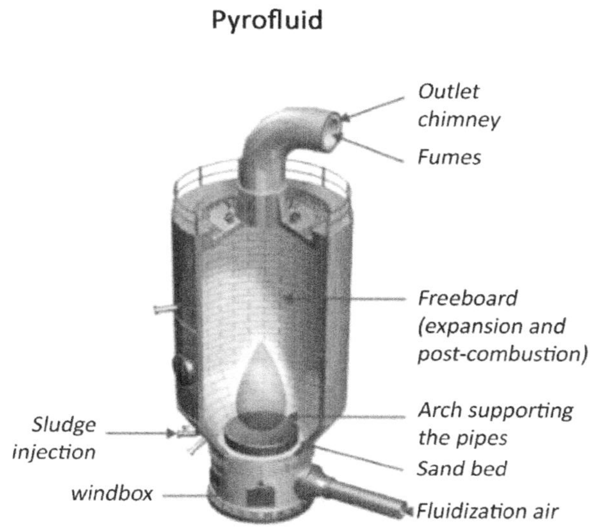

Figure 4.33. *Pyrofluid™ installation*

The heat produced in the installation can be used in a district heating network, reducing the plant's operating costs. While dedicated thermal treatment was once only feasible for large wastewater treatment plants (capacity exceeding 100,000 PE), this solution can now be considered for smaller facilities.

4.5.1.2. *Thermal hydrolysis of sludge*

Dewatering sludge from conventional anaerobic digestion is an important step in overall sludge treatment because it reduces the volume of biosolids to be treated. A digester fed with sludge that has undergone thermal hydrolysis results in sludge with better dewatering, as the thermal hydrolysis step fundamentally alters the properties of the sludge.

Dewatering sludge from conventional anaerobic digestion is an important step in overall sludge treatment because it reduces the volume of biosolids to be treated. A digester fed with sludge that has undergone thermal hydrolysis results in sludge with better dewatering, as the thermal hydrolysis step fundamentally alters the properties of the sludge. This facilitates the handling of the final product and minimizes costs, whether the sludge cake is intended for spreading, storage or incineration. Wastewater treatment plants often optimize this step by choosing the best equipment or dewatering method. However, one important factor determines sludge dewatering: the nature and characteristics of the sludge to be dewatered.

Figure 4.34. *Treatment plant with thermal hydrolysis*

4.5.1.2.1. Different sludge characteristics and different levels of sludge dewatering

A comparison between the dewatering of primary sludge and secondary activated sludge shows that there are many sludge characteristics that facilitate dewatering or lead to a higher percentage of DM.

Characteristics affecting dewatering	Primary sludge	Secondary sludge
	Dewatering difficulty →	
Viscosity	Low	High
Extracellular polymers (EPS)	Low	High
Organic or inorganic content	High: inorganic compounds	High: organic compounds
Presence of fibers	High	Low
C/N ratio	High	Low
Structure of water molecules	Free	Bound

Table 4.30. *Comparison of the characteristics of primary and secondary sludge*

These characteristics are listed in Table 4.30 and show that the difficulty of dewatering increases with viscosity and the presence of organic carbon, bound water and extracellular polymers. Such characteristics are common in activated sludge, which is known to be more difficult to dewater than primary sludge.

In addition to this difference in dewatering between primary and secondary sludge, not all equivalent substrates or raw materials have the same level of dewatering. In other words, not all primary sludge will behave the same way during dewatering, just as not all secondary or mixed sludge will be equally difficult to dewater. This is because the sludge produced at all treatment plants is generally strongly influenced by the treated wastewater, the type of treatment and the ambient temperature.

Thermal hydrolysis improves sludge dewatering by affecting the bonds between water molecules and flocs and solid matter. By modifying this characteristic, it is possible to facilitate sludge dewatering.

Water distribution refers to the water present in different states within the sludge, which determines the difficulty of separating it from the solid particles.

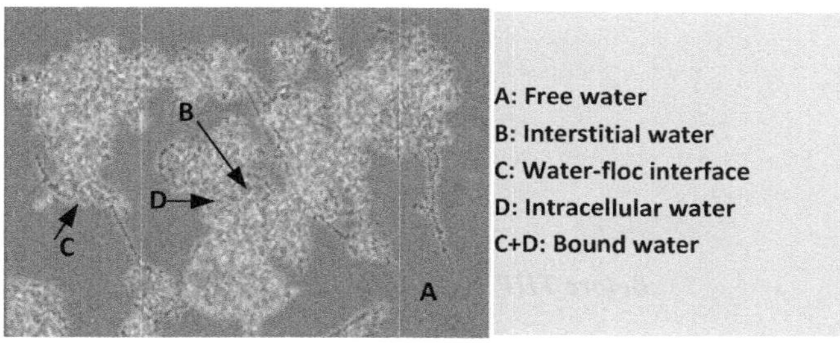

Figure 4.35. *Water distribution in a bacterial floc (Kopp and Dichtl 2001)*

The different types of water in sewage sludge are distinguished primarily by the type and intensity of their physical bonding to solids (Figure 4.35).

In a sewage sludge suspension, there are four different types of water:

– Free water that circulates freely between the sludge flocs. Not bound to the particles, it can be mechanically separated.

– Interstitial water is bound by capillary forces between the sludge flocs. The molecules within the flocs are not tightly bound to the solids.

– Water interfacing with the flocs by adhesion.

– Intracellular water that is bound by capillary forces between and within the flocs.

Interstitial water, both at the interface and within cells, is called "bound water" and cannot be removed by mechanical dewatering. Only free water can be separated during mechanical dewatering, but to eliminate bound water, a drying process is necessary.

The binding energy of free water is less than 0.28 $kJ.kg^{-1}$. The binding energy of bound water (sum of surface water and intracellular water) is greater than 5 $kJ.kg^{-1}$.

Polymer conditioning increases the rate water is released from the sludge, but the free water content is not influenced by this process.

This is where the thermal hydrolysis process (THP) can make a difference. It acts on water bonds to improve dewatering. It involves treating sludge at high temperature and pressure (160–180°C and under approximately 6 bar for about 20–30 minutes). This operation disintegrates the sludge particles and releases the interstitial water, which becomes free water that can then be mechanically separated (Figure 4.36). Releasing more water from the sludge increases its solids content and has positive effects on the operating costs of sewage sludge management facilities that must transport this final product for storage, spreading or incineration.

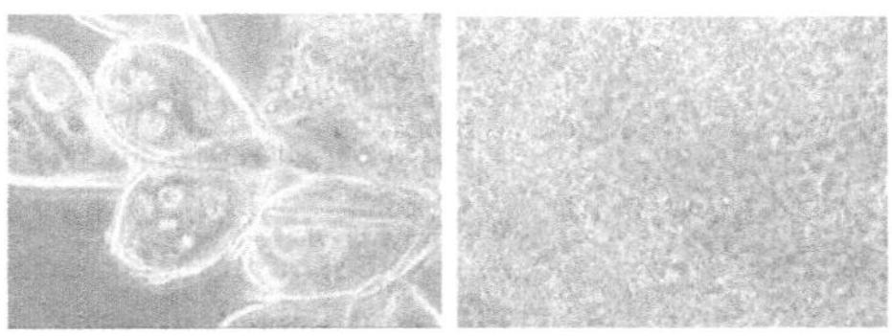

Before THP ***After THP***

Figure 4.36. *Bacterial flocs before and after THP (Svensson 2022)*

Dewatering can be assessed in two distinct ways: by the speed at which the product can be dewatered or by the final dryness that can be achieved. Because thermal hydrolysis breaks down OM and releases interstitial water, it significantly improves the dryness of the final cake. This process considerably reduces the volume of biosolids, resulting in operational savings for wastewater treatment plants.

In short, thermal hydrolysis simplifies sludge management and disposal in several ways:

– by increasing biogas production;

– by requiring less energy for the final dewatering stage;

– by significantly reducing the volume of sludge while thereby lowering handling costs;

– by destroying pathogens;

– by eliminating odors and facilitating storage and transport.

There are many THPs. Only two of them are presented here.

4.5.1.2.2. Cambi THP® process

This THP, patented by the Norwegian company Cambi, allows the cellular structure of bacteria to disintegrate into an easily digestible and dewaterable product.

Raw sewage sludge is collected during the primary and secondary treatment stages. It is then dewatered until it reaches 16–18% DM.

These thickened sludges are continuously fed into a reactor that homogenizes and preheats them to approximately 100°C. Steam recovered from the expansion reactor is used for heating. The hot sludge is then continuously conveyed to the reactors. When a reactor is full, the sludge flows to the next reactor. Typically, there are two to five reactors in a Cambi THP®, depending on the sludge volume, reactor size and hydraulic retention time (Figure 4.37).

Once the reactor is filled and sealed, steam heats the contents to 160–180°C at a pressure of approximately 6 bar. The THP typically lasts 20–30 minutes. This ensures that all pathogens are killed.

The sterilized sludge is then directed to the pressure-reduction chamber, which operates at atmospheric pressure. The sudden drop in pressure results in significant cellular destruction of the OM. It is the steam generated by the pressure release that preheats the sludge. After leaving the expansion chamber, the sludge is cooled in heat exchangers to the temperature required for anaerobic digestion. It is then sent to the anaerobic digesters.

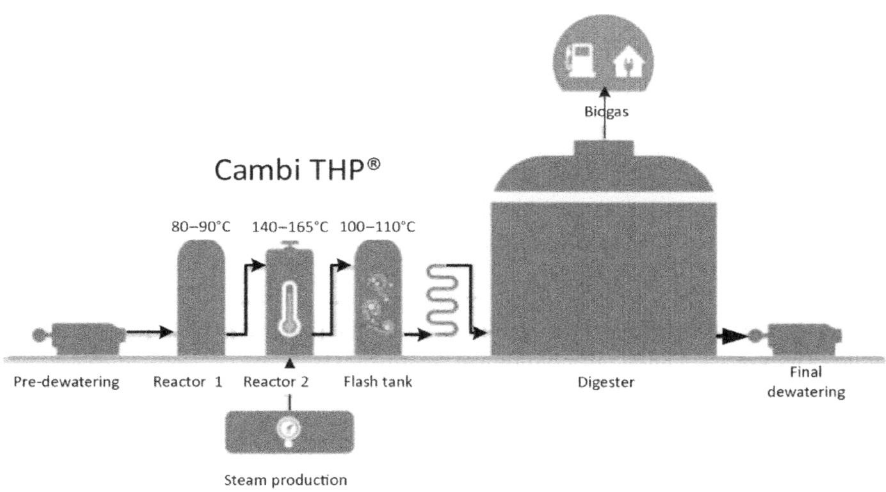

Figure 4.37. *Cambi THP® process*

4.5.1.2.3. Biothelys® (Veolia) process

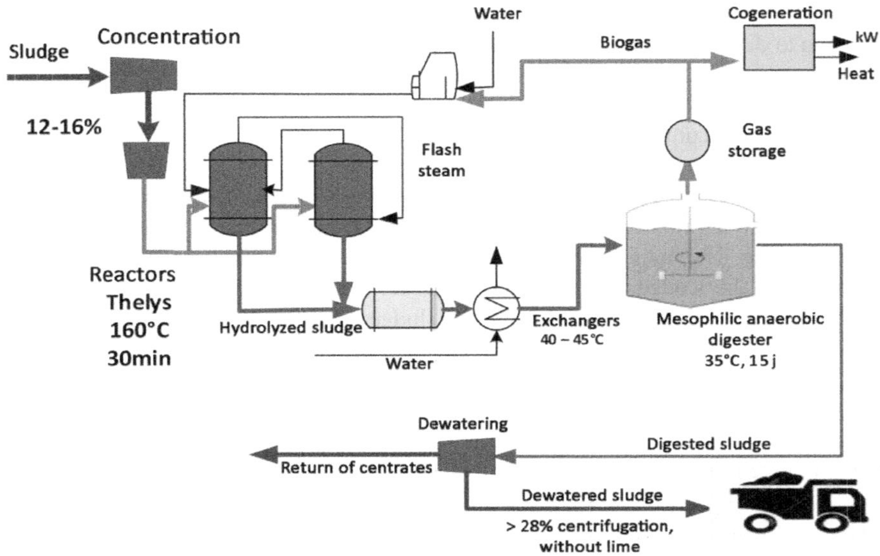

Figure 4.38. *Diagram of the Biothelys® process operation*

Anaerobic digestion has emerged as a promising method for producing green energy and reducing the mass of municipal sludge produced by wastewater treatment plants.

In recent years, Veolia developed a THP called Biothelys® to enhance the value of mesophilic anaerobic methanation of municipal or industrial sludge. Thermal hydrolysis of the dewatered sludge is carried out by injecting steam at a temperature of 160°C for approximately 30 minutes. The THP reactors are coupled to recover the flash steam and heat the sludge. Anaerobic digestion of the hydrolyzed sludge takes place over 15 days and reduces volatile matter in biological sludge by more than 45% with extended aeration (Figure 4.38).

The Biothelys® process is a combination of thermal hydrolysis of wastewater sludge (Thelys) associated with biological treatment (Bio).

4.5.1.2.4. Operating conditions

Thermal hydrolysis occurs:

– on concentrated sludge;

– at 150–180°C;

– at 8–10 bars;

– by injecting steam, produced by converting biogas from digestion;

– with a retention time of 30–60 min.

The aim of the process is to solubilize a portion of the SS into soluble materials that are more readily biodegradable. The greater biodegradability of hydrolyzed sludge compared to untreated raw sludge allows for improved performance in removing volatile matter during the downstream biological treatment stage. Thus, for anaerobic systems, biogas yield and the release of the energy potential of sewage sludge are significantly improved. The amount of sludge to be disposed of after digestion is smaller, and transport and disposal costs are reduced. Another effect of thermal hydrolysis is the decrease in sludge viscosity, which allows an anaerobic digester to be fed with sludge with a dry solids content of 10–12%.

Thermal hydrolysis operates with concentrated sludge at a dry solids content of up to approximately 15%. Initially, sludge concentration is necessary to minimize heating requirements. This step can be carried out by centrifugation, belt filtering or any other equipment producing 12–16% dry solids content.

The dewatered sludge is then stored in a hopper that feeds concentrated sludge to the reactor. The Biothelys® system consists of lines of two or three thermal hydrolysis reactors operating in parallel. All reactors have exactly the same functions, and their operation is interdependent. Each reactor operates in batch mode, and a complete cycle can last from 120–165 minutes, depending on the configuration.

A complete cycle is as follows: the first step is filling the reactor by transferring raw sludge from the hopper to the reactor using the feed pump. The second step is preheating the raw sludge with recycled steam from another reactor. The third step is steam injection to heat the sludge to the required temperature. If steam is already produced on site at 10 bar (as is the case at some industrial sites), it can be used for thermal hydrolysis; otherwise, fresh steam is produced by a dedicated steam generator or by a cogeneration unit (combined heat and power generation system).

Recycled steam and fresh steam are injected into the bottom of each reactor through specific injectors. Next comes the fourth stage, the temperature/retention time reaction: the raw sludge is exposed to the required temperature under pressure (approximately saturated steam pressure) and undergoes hydrolysis. Following the retention period, the fifth stage begins: the release of steam from one reactor to another (which is in the preheating phase). This steam release, commonly called a “flash”, has two main advantages: it allows for steam recycling and therefore energy recovery, and it cools the hydrolyzed sludge by reducing the pressure. The final stage is draining the reactor, using the residual pressure to facilitate the transfer to the buffer tank.

The final dewatering of the digested sludge is carried out using a centrifuge. The sludge dryness should be greater than 28%. The dewatered sludge is then stored in two 8-month storage cells. This storage area is located within the plant facilities. The digested sludge is of high quality: it is free of pathogens (helminth eggs, *E. coli*, enteroviruses and salmonella).

Table 4.31 compares the technical characteristics of digestion via the Biothelys® process and mesophilic and thermophilic digestion (Chauzy et al. 2008).

The Biothelys® process is an excellent disinfection system (temperature above 140°C and a minimum retention time of 20 minutes). It is also stabilized and therefore odorless, very easy to handle and stack and has excellent agricultural value.

	Biothelys® mesophilic digester	**Mesophilic digester**	**Thermophilic digester**
Applied load (kg DM.d^{-1})	3,830	3,830	3,830
Digester inlet concentration (g DM.L^{-1})	130	50	50
Retention time (d)	15	15–21	12
Digester usable volume (m^3)	435	1,608	1,608
Volumetric load (kg VM.m^{-3}.d^{-1})	5.3 (3–7)	1.4 (1–3)	2.5 (2–5)

Table 4.31. *Comparison of technical characteristics between Biothelys® digestion and mesophilic and thermophilic digestion (Chauzy et al. 2008)*

The dryness of the digested sludge should be greater than 28% without any injection or addition of lime. The first test point, with a dryness of 27%, was carried out 45 days after the digester started up. Indeed, the centrifuge reached 29% with digested sludge without lime.

The significant mass reduction combined with the good dryness achieved for the digested sludge results in a substantial reduction in the volume of wet sludge requiring disposal: 44% less wet sludge compared to conventional methanization. Methanization is generally poorly suited to the treatment of biological sludge with low solids removal (25–35%) and to biogas production. With thermal hydrolysis, the prospect of using anaerobic digestion becomes very attractive for such installations, with a removal of up to 50–55% of the residual biomass, significantly improved dewatering, and consequently, a considerable reduction in the volume of sludge requiring disposal. The energy recovery (biogas) from the biological sludge is thus optimized while producing high-quality sludge: free of pathogens, stabilized, odorless and with good fertilizing capacity.

Perfectly integrated into the wastewater treatment plant, the Biothelys® process:

– reduces the amount of sludge to be removed while improving its quality;

– reduces sludge volume by facilitating dewatering: final dryness of 28–35% after centrifugation and 35–50% after dewatering on an FP, without liming;

– controls disposal costs;

– helps optimize the digester size by feeding it with more concentrated sludge (>10% DM) and reducing the retention times;

– increases biogas production;

– adapts to all residual sludge recovery channels.

4.5.1.3. *Wet oxidation*

Autothermal wet oxidation (AWO) involves the thermal treatment of sludge, bypassing the dewatering and drying stages. The sludge is heated to 200–250°C under pressure (up to 50 bar) and in the presence of pure oxygen to prevent water evaporation. The retention time varies between 30 and 60 minutes. The reaction destroys up to 95% of the OM, transforming it primarily into carbon dioxide and ammonia.

This is what is obtained:

– an aqueous solution, returned to the head of the treatment plant for recycling;

– combustion gases subsequently treated in a catalytic reactor;

– a solid mineral residue containing 50% DM and less than 5% OM.

Thermal oxidation of sewage sludge is the solution that best meets the criteria for volume reduction and sanitization. Indeed, regardless of the combustion technique used, the sludge is completely mineralized, and pathogenic germs are completely destroyed. Today, there are two main principles of sludge oxidation: gaseous oxidation, in other words, incineration, and liquid-phase oxidation or wet oxidation.

The AWO process does not use fuel oil or any other energy source during the operation, which treats thickened sludge (4–10% dry solids content) and produces a final mineral by product called "technosand" that can be reused in various industries.

4.5.1.3.1. Athos™ process

This is a continuous process using a fully mixed reactor (1-hour drying time, 250°C and 50 bar) in which OM is oxidized in the liquid phase (for incineration, oxidation is carried out in the gaseous phase by injection of pure oxygen) (Chauzy et al. 2009).

By using a stoichiometric oxygen injection rate relative to the BOD_5 of the sludge being treated, BOD_5 removal is close to 85%. The residual BOD_5 present in the settling tank supernatant and in the FP filtrate (final dewatering stage) is readily biodegradable at over 90%. The nitrogen in the sludge to be treated is 70% transformed into $NH4^+$, the remainder being mainly soluble organic nitrogen and gaseous nitrogen (N_2). The dryness of the technosand is between 50% and 70% for a conventional plate FP and greater than 85% for a drying plate FP. The TOC content of the technosand is always less than 3% and the volatile matter content is less than 7%. The technosand comprises 34% illitic clay, 20% quartz, 20% calcium carbonate, 10% calcium phosphate and 3% ferric oxide.

The AWO process generates a fine-grained mineral by product (D50 = 2 μm), very useful to the ceramics industry (Filhol et al. 2010) due to its high content of clay elements, quartz, phosphate and calcium carbonate.

a) *b)*

Figure 4.39. *a) AWO-Athos™; b) exchangers*

The liquids generated by the process are returned to the plant inlet. When installed downstream of the thermal hydrolysis and anaerobic digestion stages, it ensures the long-term viability of the installations by guaranteeing precise control of costs and risks. This technique allows for the treatment of sludge directly from the thickener (dryness level less than 10%).

It also offers the advantages of simplifying air treatment compared to incineration (halving CO_2 emissions and producing smoke free of dust, heavy metals and dioxins) and is well-suited to medium-capacity plants (50,000–150,000 IE).

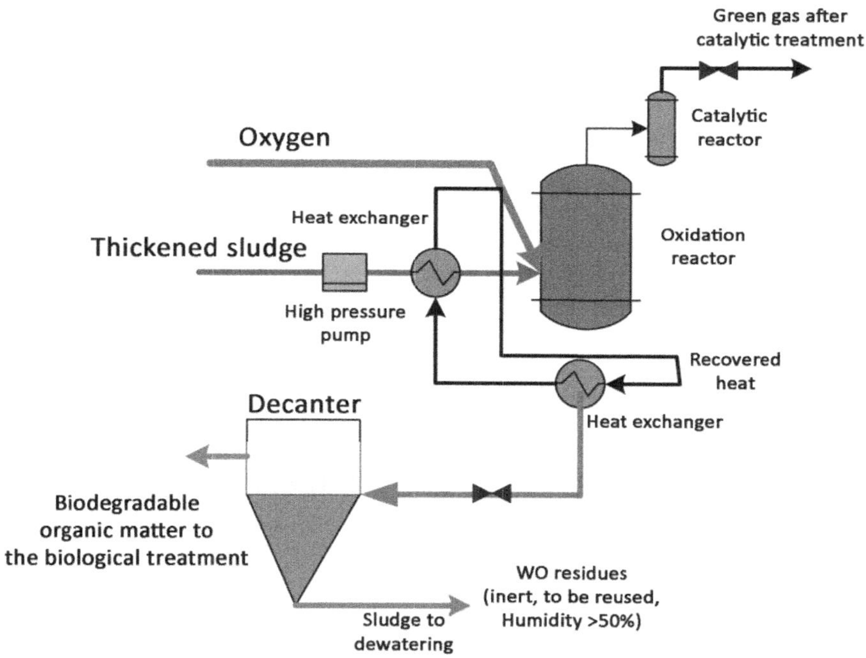

Figure 4.40. *Diagram of the Athos™ (Veolia) wet oxidation process*

4.5.2. *Other by products of sludge recovery*

4.5.2.1. *Struvite*

Struvite is a white crystalline substance composed of magnesium, ammonium and phosphate in equal molar concentrations ($MgNH_4PO_4{,}6H_2O$).

The overall formation reaction is:

$$Mg^{2+} + NH_4^{+} + PO_4^{3-} + 6H_2O \rightarrow MgNH_4PO_4{,}6H_2O$$

Struvite has a molar ratio of $Mg^{2+}/PO_4^{3-}/NH_4^+ = 1/1/1$.

Struvite precipitation requires that the various constituents be present in the correct proportions. Very often, struvite precipitates spontaneously in pipes, causing blockages and posing a problem in wastewater treatment plants. The precipitation of struvite allows for the removal of phosphorus, some nitrogen (in ammoniacal form) and magnesium. The formation and development of struvite crystals follow two chemical steps: nucleation and growth (Jones 2002).

The study of these mechanisms is complex because they involve a combination of the following factors.

4.5.2.1.1. Mass transfer phenomena: reaction kinetics

Several physicochemical parameters influence the reaction kinetics, such as the pH of the solution in which struvite can precipitate (Bouropoulos and Koutsoukos 2000), supersaturation (Doyle and Parsons 2002), mixing energy (Ohlinger et al. 1999), temperature (Aage et al. 1997), the presence of "foreign" ions (Le Corre et al. 2005) and mass transfers (Mullin 2002; Jones 2002).

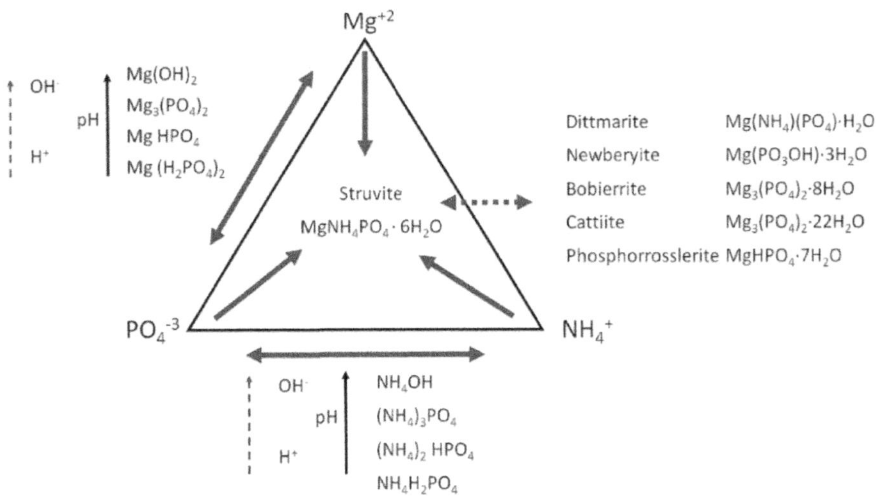

Figure 4.41. *Mechanism of struvite formation (Tansel et al. 2018)*

Over 90% of the dissolved phosphorus can be recovered from the supernatant of anaerobic digesters, and struvite allows for the recovery of between 50 and 80% of the total phosphorus (Yoshino et al. 2003; Wu and Bishop 2004). Due to its low solubility at neutral pH, struvite is an excellent slow-release fertilizer and does not cause root burn, unlike traditional ammonium phosphate-based fertilizers.

The formation of struvite requires the addition of Mg^{2+}. In the case where calcium and magnesium are present together in solution, the Ca/Mg ratio influences the type of precipitate: calcium phosphate is formed in the case of an excess of Ca^{2+}, and precipitation of magnesium phosphate in the case of excess Mg. An excess of NH_4^+ ions is very favorable to struvite precipitation and phosphorus removal (Stratful et al. 2001).

4.5.2.2. *Struvite precipitation processes*

Struvite precipitation occurs at a pH ranging from 7.5 to 10, while magnesium phosphate precipitates in a pH range of 8.5–11.

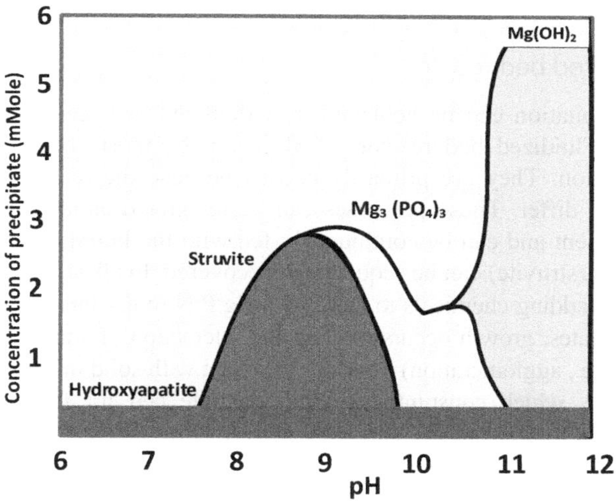

Figure 4.42. *Modeling struvite precipitation as a function of pH (Lee et al. 2003). For a color version of this figure, see www.iste.co.uk/gaid/wastewaterengineering4.zip*

Phosphorus and nitrogen are usually present in wastewater, but magnesium can be the limiting factor. This is why adding this element is necessary.

The most economical solution is Mg hydroxide ($Mg(OH)_2$) even though Mg chloride ($MgCl_2$) is more efficient.

Recovery technologies are based on the precipitation of phosphate in the form of hydroxyapatite (HAP) or struvite from the supernatants of anaerobic digesters.

The technologies can be implemented in a stirred reactor (Stratful et al. 2004; Laridi et al. 2005), in a fluidized bed reactor or even in a bubble column reactor (Ueno and Fujii 2001; Von Münch and Barr 2001; Battistoni et al. 2005).

The fluidized bed reactor is the most studied process for struvite precipitation, with varying retention times for solids and water depending on the recommended solutions. Struvite precipitates in the reactor after the addition of $MgCl_2$ to maintain the minimum molar ratio of Mg:P equal to 1:1.

The pH required for nucleation is adjusted by adding NaOH, and a helical stirrer is used to mix the struvite crystals and promote their formation. A settling zone is integrated into the reactor to collect the particles, or simple timer-controlled pumping is performed. This latter method requires a final settling step to recover the product. The supernatant is either discarded or partially recycled to accelerate nucleation.

4.5.2.2.1. Fluidized bed reactor

Struvite precipitation can be achieved in a fluidized bed reactor or in a bubble column reactor. Fluidized bed reactors (FBRs) are the most studied processes for struvite precipitation. They are primarily chosen because the retention times of the solids and liquids differ. These processes can be integrated as an additional step in wastewater treatment and can be continuously fed with the liquid to be treated, while the resulting solid (struvite) can be sequentially recovered. In all cases, struvite particles can precipitate by adding chemicals to achieve a Mg:P:N molar ratio of 1:1:1. Once the first particle nucleates, growth occurs through the interaction of small struvite particles with each other (i.e., agglomeration) or through contact with solid materials such as sand or simply struvite, which constitute an initial particle bed for fluidized beds. This solution is identical to the catalytic softening performed in potable water, during which the crystals of $CaCO_3$ attach themselves to sand particles or calcium carbonate particles.

Phosphorus recovery rates obtained with these processes can vary from 60% to 94%. The solid phase is partially or completely recovered in the reactor when the particles have reached a sufficient size.

The residence time is less than 30 minutes when the bed is well inoculated. However, the initial germination of nucleating bacteria may be slow.

Since one mole of nitrogen is equivalent to one mole of struvite in the molecular formula ($MgNH_4PO_4{\cdot}6H_2O$), the quantity of struvite can be obtained from the relationship:

$$\%\ \text{struvite} = 100\,\frac{\text{n N x M struvite}}{\text{mprecipitated}}$$

where nN is the number of moles of nitrogen in the effluent, M is the molar mass of struvite (245.41), and m precipitate is the mass of struvite produced.

Fluidized beds allow the production of crystals with a size of 1–3 mm.

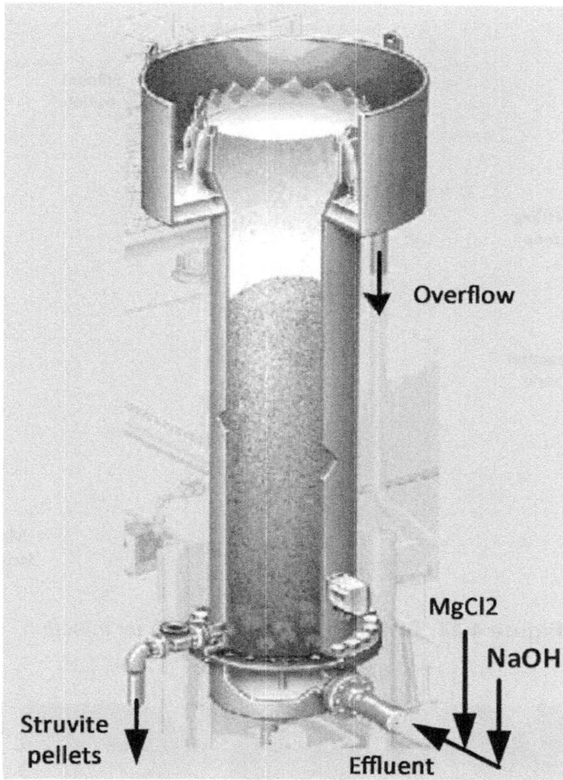

Figure 4.43. *Fluidized bed reactor*

The flow velocities (fluidization velocity) must be high (> 50 $m.h^{-1}$) due to the density of the solid (1.7 $g.cm^{-3}$), which has the risk of settling very quickly and accumulating at the bottom of the reactor. The growth of struvite crystals induces particle classification within the reactor. The largest granules accumulate at the bottom of the reactor and can subsequently be withdrawn.

Fine struvite particles are often produced during the crystallization process. Due to the high flow velocity, these fine particles can escape from the reactor. This is why the seeding velocity must be slower than the applied velocity. Another solution is to induce siltation in the upper part of the reactor to reduce the fluidization velocity and prevent fine struvite particles from exiting with the effluent. Increasing the size of the struvite particles can be achieved by injecting metallic salts (Fe^{3+}, Al^{3+}) and with a polymer using metallic salts.

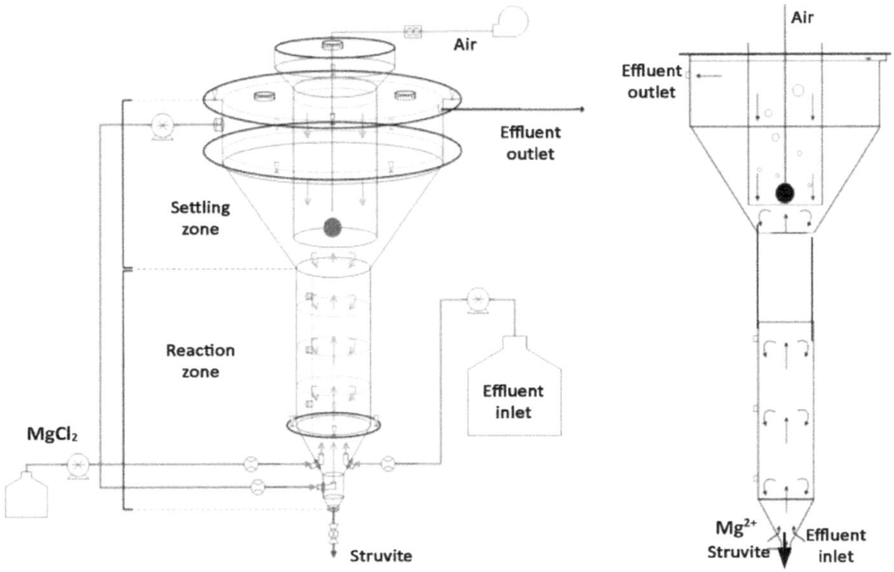

Figure 4.44. *Fluidized bed reactor with air injection*

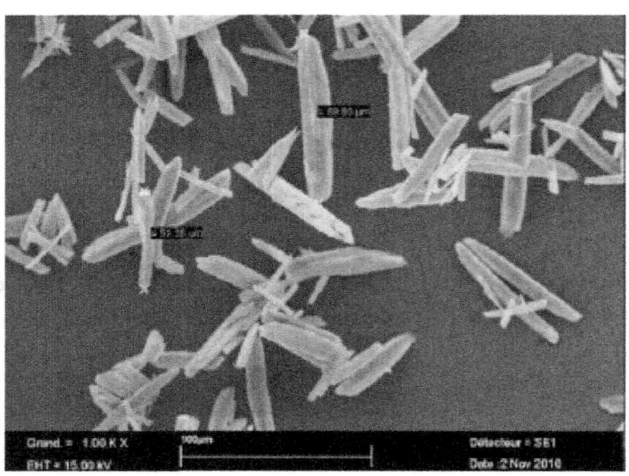

Figure 4.45. *Struvite crystals (Hanhoun 2011)*

The efficiency of phosphate recovery by struvite crystallization, as well as the struvite nucleation and growth phenomena, is fairly well known (Bouropoulos and Koutsoukos 2000; Von Münch and Barr 2001; Doyle et al. 2004; Stratful et al. 2004; Battistoni et al. 2005; Le Corre et al. 2005).

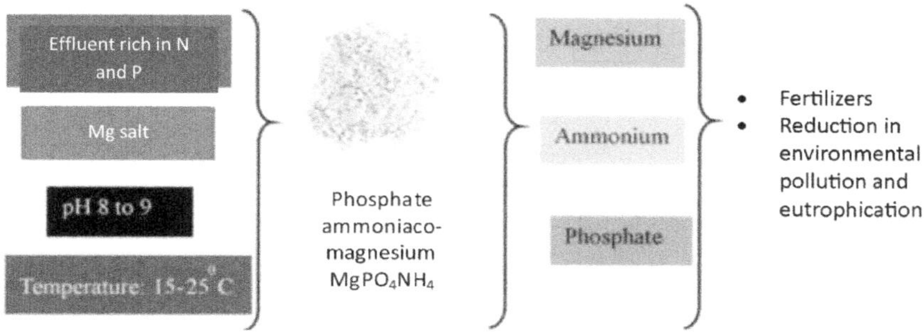

Figure 4.46. *Struvite production flow and environmental benefit*

Implementation of the struvite precipitation processes on an industrial scale remains limited. Several factors can explain this situation: poorly controlled process cost and product economic value and the need for pH control.

4.5.2.3. *Bioplastics*

Certain bacteria present in wastewater treatment processes can produce biopolymers. This property opens the way to a new solution for sewage sludge reuse.

Pens, trash bags and bank cards – the plastic used in many everyday products could soon become more sustainable. Researchers are developing a method to produce polyhydroxyalkanoates (PHAs), a type of biopolymer, from sludge generated by wastewater treatment plants. The process relies on bacteria naturally present in treatment systems. These microorganisms are selected through alternating cycles of feeding and nutrient deprivation. When nutrients become scarce, the bacteria respond by producing PHAs, which serve as internal carbon and energy reserves.

During the growth phase, substrates are carefully controlled rather than supplied randomly, as they directly influence the properties of the resulting bioplastic. Wastewater treatment by products such as sludge and grease residues are used as feedstock. The organic carbon they contain is first converted into VFAs. The specific types of VFAs produced determine the structure of the resulting PHAs, thereby influencing the mechanical and physical characteristics of the final bioplastic.

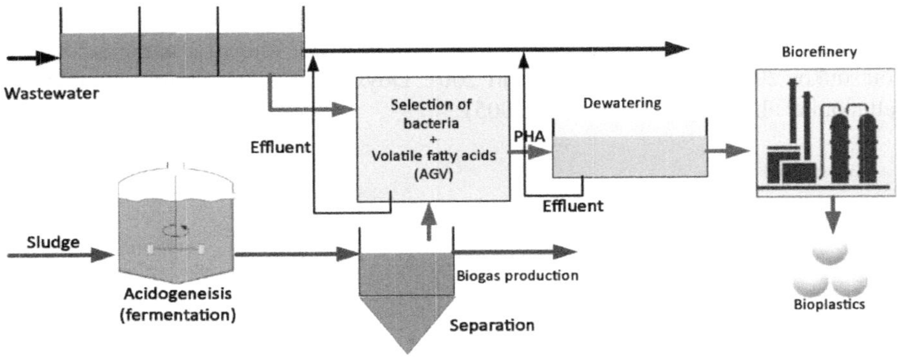

Figure 4.47. *Simplified diagram of bioplastic production*

Combining biogas and bioplastics is another approach. Indeed, a treatment plant using this process can direct part of its sludge to produce the energy (in the form of biogas) it needs to operate and the other portion to the manufacture of bioplastics. A wastewater treatment plant serving a city of 100,000 inhabitants should be able to produce approximately 500 tons of biopolymer per year, according to Veolia. Currently, 0.15 kg of PHA is obtained from one kilogram of BOD_5, but this is below the potential of 0.45 kg.

However, the social acceptability of the plastic origin may be questioned. Currently, the uses of PHA exclude contact with food. Companies should, however, be mindful of the manufacturing cost. It will also be necessary to determine the production cost, which should be lower than that obtained from current resources.

Local authorities are also interested because they see an opportunity to improve wastewater management through this solution.

4.5.3. *Conclusions on reuse*

Sludge recovery therefore enables both the direct reuse of sludge extracted from biological treatment processes and the indirect reuse of associated by products.

4.5.3.1. *Direct sludge reuse*

Treated sludge intended for spreading can be:

– Liquid sludge: dry solids content 2–5%, produced by small treatment plants (< 2,000 PE). This sludge has undergone thickening or silo storage.

– Pasty sludge: dry solids content 15–25%, resulting from biological digestion, corresponding to medium-sized treatment plants (65,000–200,000 PE). This sludge has undergone dewatering using a belt filter or centrifugation.

– Limed sludge: DM content 25–30%, this type of sludge can be pasty or solid and is produced by medium to large-sized treatment plants (20,000–100,000 PE). This sludge results from liming following dewatering.

– Composted sludge: dryness 35–70%, resulting from the composting of dewatered sludge.

– Lagoon sludge: dryness of 10–20%. This corresponds to a particular category of liquid sludge extensively treated by macrophytes.

– Solid sludge: results from FP treatment or thermal drying. It is produced mainly by large treatment plants.

Figure 4.48. *Agricultural spreading*

There are two main uses derived from agricultural spreading: revegetation and energy crops. Revegetation is still underutilized and has significant potential. Instead of using treated sludge as a soil amendment for crops, the sludge is used as a nutrient-rich substrate on inert, eroded or sparsely vegetated soils. This technique is recommended for the landscape rehabilitation of barren, cleared sites such as quarries, landfills, highway construction sites or earthworks. The development of urban green spaces is also a possibility.

Composting is first and foremost a process for stabilizing OM, then a process for organic recovery, producing compost subject to the same regulatory constraints as agricultural spreading. However, composting offers several advantages over spreading after liming:

– reduction in sludge volume and water content;

– odor reduction;

– improved handling (better structure than non-composted sludge);

– natural stabilization and sanitization without chemical additives, thus offering greater agronomic interest (a large quantity of humus rich in fertilizing elements).

Agricultural value after composting is guaranteed due to the quality of the final product.

4.5.3.2. *Indirect recovery*

Wastewater contains numerous substances that may have economic or environmental value. Solutions are being developed to transform wastewater treatment by products such as sludge into reusable raw materials or materials that can be sold to local authorities and industries:

– Transformation of OM into fertilizers (compost or struvite).

– Transformation of sludge into bioplastics: dredged sands are washed and dried and then reused as road sub-base or backfill material.

4.6. Conclusion

Wastewater treatment leads to the production of sludge containing inert and organic compounds, pollutants and pathogens. Appropriate and effective sludge treatment is therefore essential for comprehensive wastewater management.

Significant technological progress has been made in sludge treatment over the past 15 years. This involves adapting processes to a wide variety of conditions (treatment of industrial effluents at wastewater treatment plants, mixing with rainwater, presence of pharmaceutical compounds in wastewater, etc.).

In all cases, the choice of the recovery or disposal method for the sludge produced must take into account the existence and sustainability of potential outlets, particularly for spreading, and try to share existing waste treatment facilities in the area (composting platform, nearby energy recovery unit, etc.). However, French and European regulatory developments must also be taken into account, as they may create new constraints that impact the cost of sanitation.

4.7. References

Aage, H.K., Andersen, B.L., Blom, A., Jensen, I. (1997). The solubility of struvite. *Journal of Radio Analytical and Nuclear Chemistry*, 223, 213–215.

Abu-Orf, M., Junnier, R., Mah, J., Dentel, S. (2007). Demonstration of combined dewatering and thermal vacuum drying of municipal residuals. *Journal of Residuals Science and Technology*, 4(1), 25–34.

Agence de l'eau Rhône-Méditerranée Corse (2012). Méthanisation des boues de stations : règle de l'art et état des lieux sur les bassins Rhône-Méditerranée et Corse. Report, Agence de l'eau.

AMORCE (2012). Boues de station d'épuration : techniques de traitement, valorisation et élimination. Série Technique, DT 51.

Andrade, A. and Schuiling, R.D. (2001). The chemistry of struvite cristallisation. *Journal of Mineralogical*, 23, 37–46.

Andreoli, C.V., Von Sperling, M., Fernandes, F. (2007). *Sludge Treatment and Disposal*. IWA Publishing, London.

Appels, L., Baeyens, J., Degre, J., Dewil, R. (2008). Principles and potential of the anaerobic digestion of waste-activated sludge. *Progress in Energy Combustion, Sciences*, 34, 755–781.

Balmer, P. (2003). Thermal methods for sludge minimization. In *Wastewater Sludge as a Resource – Biosolids*, Trondheim, 23–25 June.

de-Bashan, L.E. and Bashan, Y. (2004). Recent advances in removing phosphorus from wastewater and its future use as fertiliser (1997–2003). *Water Research*, 38, 4222–4246.

Battistoni, P., Fava, G., Pavan, P., Musacco, A., Cecchi, F. (1973). Phosphate removal in anaerobic liquors by struvite crystallization without addition of chemicals: Preliminary results. *Water Research*, 1(11), 2925–2929.

Battistoni, P., Pavan, P., Prisciandaro, M., Cecchi, F. (2000). Struvite crystallization: A feasible and reliable way to fix phosphorus in anaerobic supernatant. *Water Research*, 34, 3033–3041.

Battistoni, P., De Angelis, A., Prisciandaro, M., Boccadoro, R., Bolzonella, D.P. (2002). Removal from anaerobic supernatants by struvite crystallisation: Long term validation and process modelling. *Water Research*, 36, 1927–1938.

Battistoni, P., Boccadoro, R., Fatone, F., Pavan, P. (2005). Auto-nucleation and crystal growth of struvite in a demonstrative fluidised bed reactor (FBR). *Environmental Technology*, 26, 975–982.

Battistoni, P., Paci, B., Fatone, F., Pavan, P. (2006). Phosphorus removal from anaerobic supernatant: Start-up and steadystate conditions of a fluidized bed reactor full-scale plant. *Industrial & Engineering Chemistry Research*, 45(2), 663–669.

Bode, H. and Klauwer, E. (1999). Advantages and disadvantages of different shapes in digester design. *WQI*, April, 35–40.

Bougrier, C., Delgenès, J.P., Carrére, H. (2007). Effects of thermal treatments on five different waste activated sludge samples solubilisation, physical properties and anaerobic digestion. *Chemical Engineering Journal*, 139(2), 236–244.

Bouropoulos, N.C. and Koutsoukos, P.G. (2000). Spontaneous precipitation of struvite from aqueous solutions. *Journal of Crystal Growth*, 213, 381–388.

Brison, C. (2004). Le séchage solaire des boues : état actuel de l'art et retours d'expérience. Report, Commission des Communautés européennes, 32.

Cabanes, F. (2006). Déphosphatation des effluents : précipitation et valorisation du phosphore. Thesis, Institut Nationale Polytechnique, University of Toulouse, Toulouse.

Carrère, H., Dumas, C., Battimelli, A., Batstone, D.J., Delgenès, J.P., Steyer, J.P., Ferrer, J. (2010). Pretreatment methods to improve sludge anaerobic degradability: A review. *Journal of Hazardous Materials*, 183, 1–15.

Chauzy, J., Graja, S., Gerardin, F., Cretenot, D., Patria, L., Fernandes, P. (2004). Minimisation of excess sludge production in a WWTP by coupling thermal hydrolysis and rapid anaerobic digestion. *Water Science & Technology*, 52(10–11), 255–263.

Chauzy, J., Cretenot, D., Bausseron, A., Deleris, S. (2008). Anaerobic digestion enhanced by thermal hydrolysis: First reference BIOTHELYS® at Saumur, France. *Water Practice and Technology*, 3(1), 12–19.

Chauzy, J., Martin, J.C., Cretenot, D., Rosiere, J.P. (2009). Wet air oxidation of municipal sludge : Return experience of the north-Brussel's waste water treatment plant. In *5th IWA Specialist Conference*, Berlin.

Cretenot, D., Chauzy, J., Patria, L., Fernandes, P. (2003). Intêret des procédés thermiques dans la problématique de la réduction et/ou de l'élimination des boues. *Tech. Sci. Méthodes*, 2, 51–56.

Davy, W. (2016). Etude de la filière boue d'une station d'épuration dans l'optique de sa modélisation globale. Master's Thesis, University of Lyon.

De Waele, O. (1929). Modelisation and complexes. *Kolloid Zeitschrift*, 47(2), 176–187.

Doyle, J.D. and Parsons, S.A. (2002). Struvite formation, control and recovery. *Water Research*, 36, 3925–3940.

Doyle, J., Philp, R., Churchley, J., Parsons, S.A. (2000). Analysis of struvite precipitation in real and synthetic liquors. *Process Safety and Environmental Protection*, 78, 480–488.

EPA (1979). Process design manual for sludge treatment and disposal. Report, Water Pollution Control Federation, Alexandria.

EPA (1987). Anaerobic sludge digestion, operations and maintenance. Report, Water Pollution Control Federation, Alexandria.

Favreau, G. (2011). Traitement et valorisation des boues de station d'épuration. Report, Amorce.

Fernández-Polanco, F. and Pérez-Elvira, S.I. (2012). Continuous thermal hydrolysis and energy integration in sludge anaerobic digestion plants. *Water Science and Technology*, 65(10), 1839–1846.

Fernández-Polanco, F., Velázquez, R., Pérez-Elvira, S.I., Casas, C., Del Barrio, D., Cantero, F.J., Fernández-Polanco, M., Rodríguez, P., Panizo, L., Serrat, J. et al. (2008). Continuous thermal hydrolysis and energy integration in sludge anaerobic digestion plants. *Water Science and Technology*, 57(8), 1221–1226.

Ferry, M. and Wiart, J. (2000). Les coûts de traitement et de recyclage agricole des boues d'épuration urbaines. *Tech. Sci. Méthodes*, 9, 117–135.

Filhol, A., Blanc, J.J., Navarro, S., Grellier, S., Cretenot, D., Kunegel, A., Prevost, J.R. (2010). Use of WAO residue in the ceramic industry. In *Qualicer'10 Conference*, Valencia.

Gaid, A. (2008). Traitement des boues. Technical Document C 5220, Techniques de l'Ingénieur.

Gilbert, A.B, Fernandes, P., Baratto, G. (2003). SAPHYR: A stabilised biosolids treatment process with flexible disposal route. In *European Biosolids & Bioresources Conference*, Wakefield.

Guibelin, É. (1999). Lutte contre la pollution des eaux – traitement des boues d'épuration. Technical Document G1451 v1, Techniques de l'Ingénieur.

Guibelin, É., Caligaris, M., Baratto, G., Fernandes, P. (2004). Procédé Saphyr Retour d'expérience d'un an d'exploitation. *Journées information eaux APTEN*, Poitiers, 29 September–1 October.

Hanhoun, M. (2011). Analyse et modélisation de la précipitation de struvite. PhD Thesis, University of Toulouse, Toulouse.

Hanhoun, M., Azzaro-Pantel, C., Biscans, B., Freche, M., Montastruc, L., Pibouleau, L., Domenech, S. (2009a). Removal of phosphate from synthetic wastewater by struvite precipitation in a stirred reactor. *COVAPHOS III*, 6, 100–106.

Hanhoun, M., Azzaro-Pantel, C., Biscans, B., Freche, M., Montastruc, L., Pibouleau, L., Domenech, S. (2009b). A thermochemical approach for struvite precipitation modelling from wastewater. In *International Conference on Nutrient Recovery from Wastewater Streams*, Vancouver.

Hanhoun, M., Montastruc, L., Azzaro-Pantel, C., Biscans, B., Freche, M., Pibouleau, L. (2011). Temperature impact assessment on struvite solubility product: A thermodynamic modeling approach. *Chemical Engineering Journal*, 167, 50–58.

Jenicek, P., Bartacek, J., Kutil, J., Zabranska, J., Dohanyos, M. (2012). Potentials and limits of anaerobic digestion of sewage sludge: Energy self-sufficient municipal wastewater treatment plant? *Water Science and Technology*, 66(6), 1277–1281.

Jones, A.G. (2002). *Crystallization Process System*. Butterworths/Heinemann, London.

Judd, S. (2023). *What Is Sludge Processing?* Judd Waste & Water Consultant, London.

Kepp, U., Machenbach, I., Weisz, N., Solheim, O.E. (1999). Enhanced stabilisation of sewage sludge through thermal hydrolysis – Three years of experience with full scale plant. *Water Science & Technology*, 42(9), 89–96.

Kopp, J. and Dichtl, N. (2001). Prediction of full-scale dewatering results of sewage sludge by the physical water distribution. *Water Science and Technology*, 43(11), 135–142.

Laridi, R., Auclair, J.-C., Benmoussa, H. (2005). Laboratory and pilot-scale phosphate and ammonium removal by controlled struvite precipitation following coagulation and flocculation of swine wastewater. *Environmental Technology*, 26, 525–536.

Le Corre, K.S., Valsami-Jones, E., Hobbs, P., Parsons, S.A. (2005). Impact of calcium on struvite crystal size, shape and purity. *Journal of Crystal Growth*, 283, 514–522.

Le Corre, K.S., Valsami-Jones, E., Hobbs, P., Parsons, S.A. (2007). Impact of reactor operation on success of struvite precipitation from synthetic liquors. *Environmental Technology*, 28, 1245–1256.

Lee, K. and Matsoukas, T. (2000). Simultaneous coagulation and breakage using constant-N Monte Carlo. *Powder Technology*, 110, 82–89.

Lee, S.I., Weon, S.Y., Lee, C.W., Koopman. B. (2003). Removal of nitrogen and phosphate from wastewater by addition of bittern. *Chemosphere*, 51, 265–271.

Liénard, A., Troesch, S., Molle, P., Esser, D. (2008a). Traitement des boues par lits plantés de roseaux : rappels des points clefs de cette technique. *Ingénieries EAT*, 41–49.

Liénard, A., Troesch, S., Molle, P., Esser, D., Thirion, F., Héritier, P., Baudez, J.C. (2008b). Valorisation des boues traitées en lits plantés de roseaux : premiers retours d'expérience des curages épandage sur quelques stations Françaises. *Ingénieries EAT*, 51–64.

Martin, G. and Laffort, P. (1991). *Odeurs et désodorisation dans l'environnement*. Lavoisier, Paris.

Metcalf & Eddy (2003). *Wastewater Engineering – Treatment and Reuse*, 4th edition. McGraw-Hill, New York.

Metcalf & Eddy (2014). *Wastewater Engineering – Treatment and Resource Recovery*, 5th edition. McGraw-Hill, New York.

Müller, J.A. (2003). Mechanical desintegration to reduce final sludge production. In *IWA Leading Sludge Conference – Drinking Water & Wastewater – Treatment Technologies*, Nordwijk and Amsterdam.

Mullin, J.W. (2002). *Crystallisation*, 4th edition. Butterworth Heinemann, Oxford.

Ndoh Rossier, M., Membrez, Y., Mottet, A. (2007). Comparaison des technologies de prétraitement des boues d'épuration pour l'augmentation de la production de biogaz. Office fédéral de l'énergie.

Nielsen, S. (2003). Sludge drying reed beds. *Water Science and Technology*, 48(5), 101–109.

Nielsen, P.H., Thomsen, T.R., Nielsen, J.L. (2004). Bacterial composition of activated sludge-importance for floc and sludge properties. *Water Sciences Technologies*, 49(10), 51–58.

Niessen, W.R. (1978). *Combustion and Incineration Processes – Applications in Environmental Engineering*. Marcel Dekker, New York.

Ødegaard, H. (2004). Sludge minimization technologies – An overview. *Water Science and Technology*, 49(19), 31.

Office international de l'eau (2001). Service national d'information et de documentation sur l'eau, (SNIDE), la stabilisation des boues de station d'épuration : techniques de mesure du procédé. Report.

Ohlinger, K.N., Young, T.M., Schroeder, E.D. (1998). Predicting struvite formation in digestion. *Water Research*, 32, 3607–3614.

Ohlinger, K.N., Young, T.M., Schroeder, E.D. (1999). Kinetics effects on preferential struvite accumulation in wastewater. *Journal of Environmental Engineering*, 125, 730–737.

Ohlinger, K.N., Young, T.M., Schroeder, E.D. (2000). Post digestion struvite precipitation using a fluidised bed reactor. *Journal of Environmental Engineering*, 126, 361–368.

Pickworth, B., Adams, J., Panter, K., Solheim, O.E. (2006). Maximizing biogas in anaerobic digestion by using engine waste heat for thermal hydrolysis pre-treatment of sludge. *Water Science & Technology*, 54(5), 101–108.

Société québécoise de phytotechnologie (2014). Fiches techniques de la SQP. 2, Les lits de séchage plantés de macrophytes [Online]. Available at: www.phytotechno.com.

Stratful, I., Scrimshaw, M.D., Lester, J.N. (2001). Conditions influencing the precipitation of magnesium ammonium phosphate. *Water Research*, 35(17), 4191–4199.

Stratful, I., Scrimshaw, M.D., Lester, J.N. (2004). Removal of struvite to prevent problems associated with its accumulation in wastewater treatment works. *Water Environment Research*, 76, 437–4199.

Svensson, K. (2022). Sludge dewatering: How does thermal hydrolysis improve dewaterability? *Cambi* [Online]. Available at: https://www.cambi.com/blog/sludge-dewatering.

Tansel, B., Lunn, G., Monie, O. (2018). Struvite formation and decomposition characteristics for ammonia and phosphorus recovery: A review of magnesium–ammonia–phosphate interactions. *Chemosphere*, 194, 504–514.

Ueno, Y. and Fujii, M. (2001). Three years experience of operating and selling recovered struvite from full-scale plant. *Environmental Technology*, 22, 1373–1381.

Van Lier, J.B., Rebac, S. Lettinga, G. (1997). High rate anaerobic wastewater treatment under psychrophilic and thermophilic conditions. *Water Science and Technology*, 35, 199–206.

Vincent, J., Molle, P., Wisniewski, C., Liénard, A. (2011). Sludge drying reed beds for septage treatment: Towards design and operation recommendations. *Bioresource Technology*, 102(17), 8327–8330.

Vincent, J., Forquet, N., Molle, P., Wisniewski, C. (2012). Mechanical and hydraulic properties of sludge deposit on sludge drying reed beds (SDRBs): Influence of sludge characteristics and loading rates. *Bioresource Technology*, 116, 161–169.

Von Münch, E. and Barr, K. (2001). Controlled crystallisation for removing phosphorus from anaerobic digester side stream. *Water Research*, 35, 151–159.

Warden, J.H. (2003). Water research center, sludge treatment plant for water work. Wastewater sludge as a resource. Technical Report, Trondheim.

Weemaes, M., Grootaerd, H., Simoens, F., Huymans, A., Verstraete, W. (2000). Ozonation of sewage sludge prior to anaerobic digestion. *Water Science and Technology*, 42(9), 175–178.

Winter, A. and Müller, J.A. (2002). Comparison of disintegration methods at full-scale anaerobic digestion plant. In *7th Latin American Workshop and Symposium on Anaerobic Digestion*, Merida, 596–602.

Wu, Q. and Bishop, P.L. (2004). Enhancing struvite crystallisation from anaerobic supernatant. *Journal of Environmental Engineering Science*, 3, 21–29.

Wu, Q., Bishop, P.L., Keener, T.C. (2005). A strategy for controlling deposit of struvite in municipal wastewater treatment plants. *Water Environment Research*, 5, 199–207.

Yoshino, M.A., Yao, M.A., Tsuno, H.B., Somiya, I.B. (2003). Removal and recovery of phosphate and ammonium as struvite from supernatant in anaerobic digestion. *Water Science and Technology*, 48, 171–178.

5

Odor Treatment

5.1. General summary

Most new wastewater treatment plants constructed worldwide are now located near residential developments. As a result, effective odor containment and treatment have become essential. Increasing housing density, heightened public sensitivity to odors and the expanded use of odor-generating processes within treatment facilities – such as sludge drying and composting – have further intensified these challenges. These processes can generate volatile organic compounds (VOCs) and other odorous substances, placing odor control at the forefront of plant design and operation. Consequently, regulatory and project requirements related to odor emissions have become more stringent. In addition to traditional limits on specific pollutants such as hydrogen sulfide (H_2S), ammonia (NH_3) and mercaptans, there is now a growing emphasis on odor nuisance itself. This is typically expressed as a maximum allowable odor concentration measured in European odor units (OUe).

Odor concentration is determined through dynamic olfactometry, based on the assessment of a trained odor panel. Odor criteria may apply directly to exhaust emissions at the discharge point or to ambient air quality in the surrounding environment. In the latter case, atmospheric dispersion modeling is required to predict the distribution of odor plumes under varying meteorological conditions, often represented as isopleth maps of odor concentration around the facility. The use of electronic noses can provide continuous, real-time monitoring to support compliance verification. Meeting these increasingly demanding objectives requires careful design and reliable operation of wastewater treatment facilities, ensuring that odor-causing compounds are properly captured and treated using appropriate control technologies.

Air pollutants are highly diverse chemical species. Some odorous substances have a fermentative origin (amines, mercaptans, aldehydes, acids, cresols). Emissions of these products are released as either diffuse (landfills) or localized (different structures within the treatment plant) sources. The concentration and flux of these chemical substances vary considerably. The relatively small molecular size and volatility of odorous compounds allow them to disperse readily into the atmosphere. Many of these compounds are water-soluble, enabling them to dissolve in the mucus lining of the nasal cavity. Once in solution, they can diffuse through the mucus and interact with the olfactory receptor cells located on the olfactory epithelium. Their electronic structure and molecular polarity enable them to bind to specific odorant receptors. This interaction triggers a transmembrane ionic current, leading to depolarization of the receptor cell and the initiation of a neural signal (Martin and Laffort 1994). These properties show that the main compounds are polar and can be nucleophilic (S, O, N), aromatic unsaturated (aldehydes, acids) or renucleophilic and unsaturated (phosphene, indole, skatole). Thus, a wide variety of VOCs, volatile sulfur compounds (VSCs) and volatile fatty acids (VFAs) can be formed and released by wastewater treatment plants. More specifically, sulfur compounds such as hydrogen sulfide (H_2S) and mercaptans, organic acids such as acetic acid, and nitrogen compounds such as ammonia (NH_3) are the main air pollutants emitted by wastewater treatment plants. Odorous compounds are generally volatile, corrosive and irritating, even at very low concentrations, and cause unpleasant odors due to their low odor threshold.

The vast majority of these substances are highly harmful and very unpleasant. Other air pollutants are poorly water-soluble compounds (halogenated derivatives, CHA, NOx) or are similar in nature to mucus (CO_3H, CO_3^{2-}, salts).

Urban wastewater, such as many types of industrial wastewater, generates foul-smelling products, a form of pollution. These products are mainly composed of nitrogen compounds (ammonia and amines, in particular) and sulfur compounds (hydrogen sulfide, mercaptans), as shown in Table 5.1.

In general, most areas of a wastewater treatment plant to some degree have odors. Depending on the site location, problems are most often encountered in the primary treatment and sludge management areas, in addition to the following areas:

– pretreatment;

– primary settling tanks;

– clarifiers;

– sludge treatment;

– storage and recovery.

Organic compounds such as aldehydes, ketones and hydrocarbons are also found in much smaller quantities, mainly in industrial gaseous effluents, whereas aldehydes and ketones are generally not found in the gaseous effluents from wastewater treatment plants.

Compound class	Compound	Formula	Olfactory threshold (mg.Nm^{-3} air)	Concentration of discharge from stations
Sulfur	Hydrogen sulfide	H_2S	0.0001–0.03	2–14
	Methyl mercaptan Ethyl mercaptan Dimethyl sulfide Diethyl sulfide Dimethyl disulfide	CH_3SH C_2H_5SH $(CH_3)_2S$ $(C_2H_5)_2S$ $(CH_3)_2S_2$	0.0005–0.08 0.0001–0.03 0.0025–0.65 0.0045–0.31 0.0030–0.014	0.5–6
Ammonia nitrogen	Ammonia	NH_3	0.5–37	5–20
Nitrogen	Methylamine Ethylamine Dimethylamine Indole Scatole Cadaverine	CH_3NH_2 $C_2H_5NH_2$ $(CH_3)_2NH$ C_8H_6NH C_9H_8NH $NH_2(CH_2)_5NH_2$	0.021 0.05–0.83 0.047–0.16 0.0006 0.0008–0.10 –	0.5–4
Acids	Acetic acid Butyric acid Valeric acid	CH_3COOH C_3H_7COOH C_4H_9COOH	0.025–6.5 0.0004–3 0.0008–1.3	Traces
Aldehydes and ketones	Formaldehyde Acetaldehyde Butyraldehyde Isovaleraldehyde Acetone	$HCHO$ CH_3CHO C_3H_7CHO $(CH_3)_2CHCH_2CHO$ CH_3COCH_3	0.033–12 0.04–1.8 0.013–15 0.072 1.1–240	Traces

Table 5.1. *Characteristics of the main compounds responsible for odors in wastewater treatment plants*

Pollutant	Formula	Perception threshold (mg.m^{-3})
Allyl disulfide	$H_2C = CH\ CH_2$ - S	1.0 10^{-4}
Allyl mercaptan	$H_2CHC\ H_2SH$	5.0. 10^{-5}
Carbon disulfide	CS_2	7.7
Dimethyl sulfide	H_3C - S – CH_3	2.0. 10^{-2}
Ethyl mercaptan	C_2H_5SH	1.6 10^{-5}
Ethyl sulfide	$C_2H_5S\ C_2H_5$	2.5. 10^{-4}
Hydrogen sulfide	H_2S	1.1 10^{-3}
Methyl mercaptan	H_3CSH	1.1. 10^{-3}
Propyl mercaptan	C_3H_7SH	7.5 10^{-5}
Sulfide dioxide	SO_2	30

Table 5.2. *Nature of sulfurous odor compounds*

In general, applicable regulations define threshold concentration values for key odor-causing compounds that correspond to conditions under which no odor nuisance is expected in nearby residential areas. Table 5.3 specifies these values.

Compound	Recommended concentration at the deodorization outlet (mg.Nm^{-3} air)
Hydrogen sulfide (H_2S) Total sulfides (in H_2S)	≤ 0.1 ≤ 0.15
Mercaptans (CH_3SH) Dimethylsulfide (CH_3SCH_3) Dimethyldisulfide ($CH_3S_2CH_3$)	≤ 0.07 of the total
Ammonia (NH_3) Amines (in CH_3NH_2)	≤ 5 ≤ 0.1

Table 5.3. *Permissible outlet concentrations*

Nitrogen and sulfur compounds are the most significant sources of pollution, and physicochemical and biological removal techniques have been implemented to combat them. While physicochemical removal techniques (adsorption, absorption, scrubbing, etc.) yield promising results, they often remain prohibitively expensive due to the high operating costs associated with the large quantities of chemical reagents required.

Depending on the different gas emissions released at wastewater treatment plants, three standard levels of air pollution can be defined:

– Low air pollution: corresponds to a fully enclosed treatment facility in which the ventilation rate is designed to remain well below the maximum allowable exposure concentration for each pollutant. This approach includes targeted containment and ventilation of the most highly emitting areas, particularly within sludge treatment and preliminary treatment processes.

– Moderately polluted to polluted air: corresponds to a covered treatment plant, with a ventilation rate calculated to comply with the maximum allowable exposure concentration of each pollutant and with containment of the most polluted points in the sludge treatment and pretreatment processes.

– Highly polluted air: corresponds to a station where only the most polluted points of the pretreatment and sludge treatment are ventilated (containment).

Beyond these values, the gaseous effluents to be treated become site-specific and may originate from the input of industrial effluents heavily polluted with nitrogen and/or sulfur compounds (rendering, for example).

Table 5.4 shows the concentrations of air heavily laden with sulfur compounds and ammonia from an industrial rendering plant.

As for aldehydes and ketones, their concentrations are generally below the nuisance threshold.

Category	Ammonia NH_3 (mg/m^3)		Nitrogen compounds (RNH_2) (mg/m^3)		Sulfur gas H_2S (mg/m^3)		Mercaptans RSH (mg/m^3)	
	Min.	**Max.**	**Min.**	**Max.**	**Min.**	**Max.**	**Min.**	**Max.**
Air with low pollution	1	10	0.5	1	2	4	0.5	1
Moderately polluted air	5	10	1	2	5	10	1	3
Polluted air	8	16	1	3	5	10	1.5	5
Highly polluted air	10	20	2	4	7	14	3	6
Air with high S level	20	20	2	6	> 30	–	> 6	–

Table 5.4. *Different categories of polluted air*

5.2. Choice of deodorization process

Selecting a deodorization process takes into account the following parameters:

– the level of guarantees to be provided;

– investment costs;

– operating costs;

– maintenance and upkeep;

– the size of the structures.

Table 5.5 compares elements of different deodorization processes.

	Adsorption on activated carbon	Autotrophic biodeodorization	Heterotrophic biodeodorization	Chemical deodorization
Guarantee and reliability	Good elimination of sulfur compounds (< 30 mg.Nm^{-3}) Problem with NH_3 in some cases	Good elimination of sulfur compounds, NH_3, amines Problem with mercaptans (> 2 mg.Nm^{-3})	Effective removal of pollutants (< 15 mg.Nm^{-3})	Very good pollutant removal depending on the number of rotations
Investment costs	Very high costs due to the price of activated carbon	Average costs (concrete, sprinkler system, air distribution)	Average costs (concrete, sprinkler system, air distribution)	High costs (towers, pumps, regulation, packing)
Operating costs	High costs (changing activated carbon)	Low costs	Low costs	Average costs (reagents, pumps)
Maintenance and upkeep	Can operate intermittently	Does not tolerate prolonged periods of inactivity	Does not tolerate prolonged periods of inactivity	pH monitoring and reagent supply Can operate in batch mode
Installation size	Compact installation Flow velocity = 720 m.h^{-1}	Compact installation Flow velocity = 500 m.h^{-1}	Fairly bulky installation Flow velocity = 100 m.h^{-1}	Large installation

Table 5.5. *Comparison of different deodorization processes*

5.3. Strategies for the biological removal of odorous products

Controlling wastewater odors is important for protecting public health and the environment. Strategies to mitigate odor nuisances at wastewater treatment plants involve monitoring, assessing and controlling odorant emissions into the atmosphere. However, this conventional approach presents significant technical and analytical challenges because odors are complex gaseous mixtures at very low concentrations. Odor pollution in wastewater treatment plants also significantly varies over time, which can be linked to weather conditions, effluent load characteristics and variations in operational conditions during wastewater treatment. In this context, odor control strategies are evolving towards an approach integrated with wastewater treatment processes. A wide range of physicochemical and biological odor control technologies are available to mitigate odor nuisances from these plants.

a) b)

Figure 5.1. *Sources of odor emissions: a) inlet of raw wastewater and b) primary settling tank*

There are two main types of wastewater odor control: liquid phase and vapor phase. Liquid phase odor control involves treating the wastewater itself to prevent the formation of odor compounds. This can be done by adding oxidizing chemical reagents, such as chlorine, hydrogen peroxide or ozone. Vapor phase odor control involves capturing and treating odorous gases released by the wastewater treatment process. The best method for wastewater odor control varies depending on the specific source of the odor and environmental conditions.

In addition to the methods mentioned above, there are a few other measures that can be taken to control wastewater odors:

– Properly maintain the wastewater treatment plant to prevent leaks and spills.

– Cover open structures and extract the air for treatment.

By using the right methods, wastewater treatment plants can effectively control odors and prevent them from becoming a nuisance.

Figure 5.2. *Example of source of odor emission: primary settler and clarification*

In wastewater treatment plants, there are numerous remediation techniques, either of a physicochemical or biological nature.

5.3.1. *Physicochemical treatment*

Chemicals with a specific pH are applied to the air extraction points of various odor sources at the treatment plant. Physicochemical treatment by physical absorption is achieved by adding chemical reagents to induce the transfer of pollutants from the gaseous phase to the liquid phase.

Chemical scrubbing towers are widely used and highly effective at removing hydrogen sulfide, ammonia and their derivatives. This chemical technology is most commonly used for high air flow rates and polluted air concentrations. Chemical scrubbers provide relatively short gas residence times; however, they typically involve higher long-term operating and maintenance costs. These costs are primarily driven by ongoing chemical consumption, scaling and fouling within the scrubbing system that require periodic acid cleaning and routine mechanical maintenance. This chemical solution requires qualified maintenance personnel dedicated to the operation and upkeep of the equipment. It is also better suited to intermittent loads than a biological system.

Based on these elements, the design of chemical deodorization plants is most often based on a configuration with two, three or four scrubbing towers:

– The first tower is acidic. It operates at pH 3, obtained by injecting sulfuric acid (H_2SO_4). This tower eliminates all nitrogen compounds (NH_3 and amines). It is necessary to prevent the formation of chloramines in the following tower(s) and thus avoid unpleasant odors outside the plant.

– The second and third towers are the basic and oxidizing towers. These towers eliminate all sulfur compounds in a basic medium and in the presence of chlorine (H_2S, RSH, etc.). The first operates at pH 8.8 (addition of NaOH) to remove H_2S. The second operates at pH 11 (addition of NaOH and chlorine) to remove R-SH. The active chlorine concentration in the tank bottom ranges from 0.5 to 1 $g.L^{-1}$ of Cl_2. Oxidation is achieved by adding bleach or by producing active chlorine through electrochlorination.

– The fourth tower is slightly basic and reducing. It operates at pH 7. Sodium bisulfite is used as the reducing agent. This tower removes aldehydes, acids, ketones and residual chlorine.

Depending on the different air categories, one acid tower (pH 3), two acid and basic towers (pH 3 and pH 11) or three scrubbing towers (pH 3, pH 8 and pH 11) are necessary to reach the expected thresholds.

However, a chemical deodorization system consisting of only a single scrubbing tower at pH 11 (H_2S removal) must be assessed with great caution. The ammonia present in the gaseous effluent would react with hypochlorite to form chloramines, the diffusion of which into the air would cause health hazards.

Air–water–chemical scrubbing towers neutralize chemical compounds and limit odor release. Naturally, the specific chemical product, and especially the specific pH, depends on the nature of the compounds present in the polluted air.

To minimize noise, the fans are installed in a soundproofed room within the installation. At the base of the tower, a recirculation pump draws water from the scrubbing baths to ensure the gas to be treated is sprayed above the packing.

The pumps operate according to the level measurement at the base of the tower. Thus, air circulates from bottom to top within the tower, while the washing solution circulates from top to bottom. The exchange of pollutants between air and water therefore occurs within the packing through mixing of the two phases. A demister at the outlet of each tower retains the airborne droplets.

Category	Two towers				Three towers			
	NH_3 ($mg.Nm^{-3}$) max.	Norg ($mg.Nm^{-3}$) max.	H_2S ($mg.Nm^{-3}$) max.	RSH ($mg.Nm^{-3}$) max.	NH_3 ($mg.Nm^{-3}$) max.	Norg ($mg.Nm^{-3}$) max.	H_2S ($mg.Nm^{-3}$) max.	RSH ($mg.Nm^{-3}$) max.
Air with low pollution	0.17	< 0.01	0.01	< 0.01	0.17	< 0.01	< 0.01	< 0.01
Moderately polluted air	0.17	< 0.01	0.02	< 0.01	0.17	0.01	< 0.01	< 0.01
Polluted air	0.28	0.01	0.02	< 0.01	0.28	0.01	< 0.01	0.01
Highly polluted air	0.35	0.03	0.04	< 0.01	0.35	0.03	< 0.01	0.02

Table 5.6. *Chemical deodorization performance for the most commonly encountered air categories*

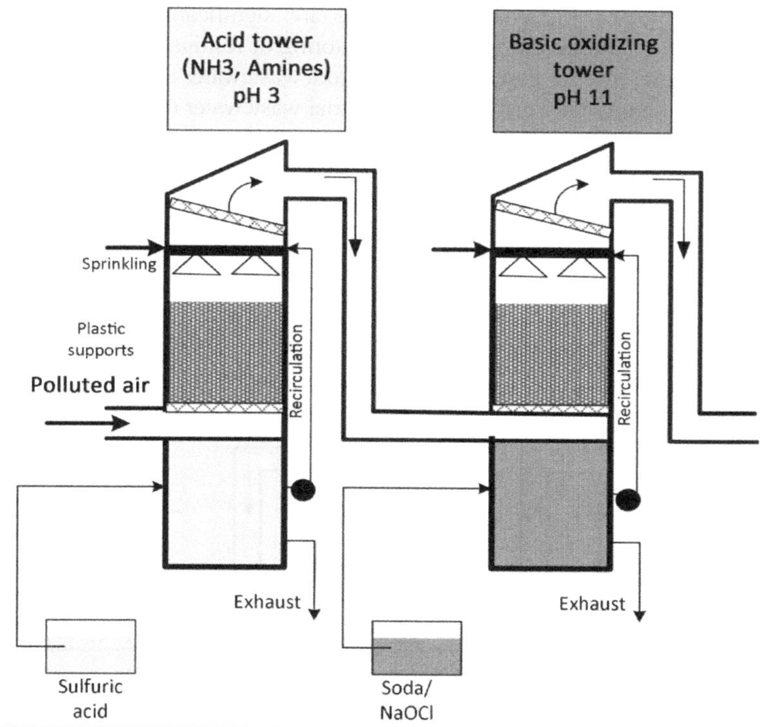

Figure 5.3. *Chemical deodorization with two scrubber towers. For a color version of this figure, see www.iste.co.uk/gaid/wastewaterengineering4.zip*

Gradually, the salts produced by the odor-removal reactions contaminate the wash-bath water. Regular blowdown is performed on the tower to control dissolved solids concentration. To compensate for the losses due to these blowdowns, additional water is required. Make-up water is supplied through a solenoid valve that is automatically actuated based on the liquid level at the bottom of the tower.

The reagent is injected into the tower using a metering pump. The quality of the washing solution, which is depleted as air passes through, is continuously monitored by a device specific to the type of reagent. This measurement controls the acid dosing pumps. In addition to this pH-based control, all the pumps are also controlled by the level in the storage tank.

When a three-tower configuration is chosen, a third scrubbing (bisulfite) tower is installed when aldehydes and ketones are present in the incoming gaseous effluent (concentrations of > 1 mg.Nm^{-3}), or if there are significant variations in sulfur compound concentrations at the inlet, then monitoring of residual chlorine is necessary – which is rarely the case for gaseous effluents from wastewater treatment plants. These compounds may be present when specific industrial wastewater (from chemical solvent industries, for example) is connected to the plant.

This third, slightly basic and reducing tower operates at pH 7. Sodium bisulfite ($NaHSO_3$) is used as a reducing agent. This tower removes aldehydes, acids, ketones and residual chlorine.

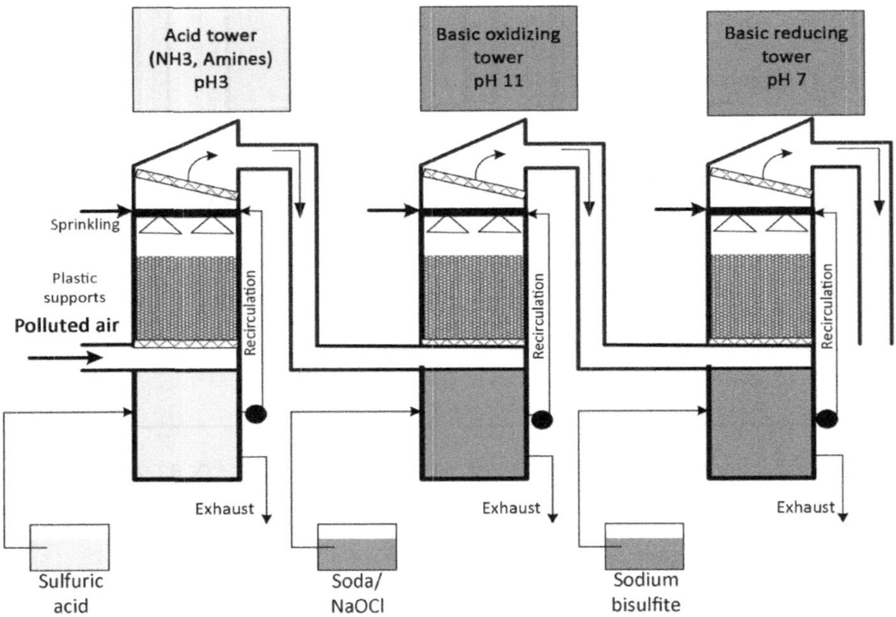

Figure 5.4. *Chemical deodorization with three scrubber towers (in the presence of aldehydes and ketones). For a color version of this figure, see www.iste.co.uk/gaid/wastewaterengineering4.zip*

5.3.1.1. *Reagent consumption in processes with two and three scrubber towers*

Using two scrubbing towers instead of three may be sufficient to meet the required thresholds for gaseous effluent emissions; choosing a single tower in a basic environment leads to higher reagent consumption.

Category	Caustic soda consumption (l/h/1,000 m^3 air)	
	Two towers	Three towers
Slightly polluted air	0.22	0.21
Moderately polluted air	0.33	0.30
Polluted air	0.35	0.33
Heavily polluted air	0.46	0.41

Table 5.7. *Caustic soda consumption according to different air categories*

5.3.1.2. *Case of gaseous effluents heavily loaded with H_2S (> 30 mg.Nm^{-3})*

When the H_2S concentration exceeds 30 mg.Nm^{-3}, the second tower can no longer reliably meet discharge concentration limits without increasing the chlorine dosage.

H_2S inlet concentration (mg.Nm^{-3})	Outlet concentration, two towers		Outlet concentration, three towers	
	H_2S (mg.Nm^{-3}) max.	RSH (mg.Nm^{-3}) max.	H_2S (mg.Nm^{-3}) max.	RSH (mg.Nm^{-3}) max.
30	0.08	< 0.01	< 0.01	< 0.01
40	0.10	< 0.01	< 0.01	< 0.01
50	0.13	< 0.01	< 0.01	0.01

Table 5.8. *Performance of two- and three-tower installations according to high H_2S concentrations*

In this case, a two-tower process can be retained by increasing the chlorine concentration, from 1 g.L^{-1} to 1.3–1.6 g.L^{-1}. The resulting performance data are shown in Table 5.9.

There will then be higher annual additional reagent costs due to the increase in the quantities of sodium hypochlorite. These values are based on outlet H_2S concentrations of 0.05 mg.Nm^{-3}. Considering all the elements involved in a chemical deodorization process, it is recommended to use two scrubbing towers when the H_2S concentration is less than 30 mg.Nm^{-3}. When the H_2S concentration is greater than 30 mg.Nm^{-3}, the same arguments apply to airflow rates below 20,000 Nm^3.h^{-1}. When these flow rates are higher, a system with three towers is recommended.

H_2S concentration ($mg.Nm^{-3}$)	Cl_2 ($g.L^{-1}$)	Two towers	
		H_2S ($mg.Nm^{-3}$) max.	RSH ($mg.Nm^{-3}$) max.
30	1	0.08	< 0.01
	1.2	0.05	< 0.01
40	1	0.10	< 0.01
	1.3	0.05	< 0.01
50	1	0.14	< 0.01
	1.5	0.05	< 0.01

Table 5.9. *Increase in treatment efficiency with a two-tower configuration as chlorine dosage rises*

5.3.1.3. *Recommended device for maintaining tower levels*

The reagent solution is stored in the tank base. From there, the following operations are carried out: pumping to the spray system, automatic deconcentrating (purging), topping up with water and pure reagent, draining and overflow. The system must have two levels:

– A low-level condition detected in the sump protects the pump from running dry.

– A high-level overflow setpoint is included to protect against overfilling of the tank in the event of blowdown system failure or uncontrolled softened make-up water flow to the top of the tower.

Softened water and chemical reagents are introduced just upstream of the reagent recirculation pump, thereby ensuring a continuous overflow at the tank bottom. This overflow would be continuously discharged without requiring any special regulation, as it is dependent on the softened water supply flow rate.

This type of treatment eliminates all malodorous compounds in a small area. This is why chemical treatment using one to four chemical scrubbers is widely implemented. However, the cost of the chemicals (sulfuric acid, sodium hydroxide and hypochlorite) required to maintain performance represents a significant expense, not to mention the associated electricity consumption for recirculating the adsorbent solutions. When comparing these operating costs, a biological treatment can be more economical and more environmentally sustainable.

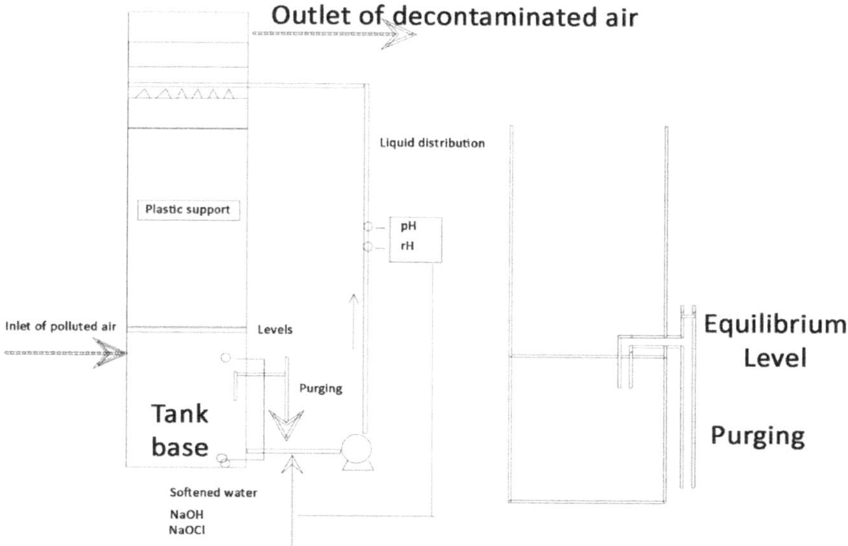

Figure 5.5. *Chemical tower purging system*

Figure 5.6. *Chemical odor treatment with three scrubber towers*

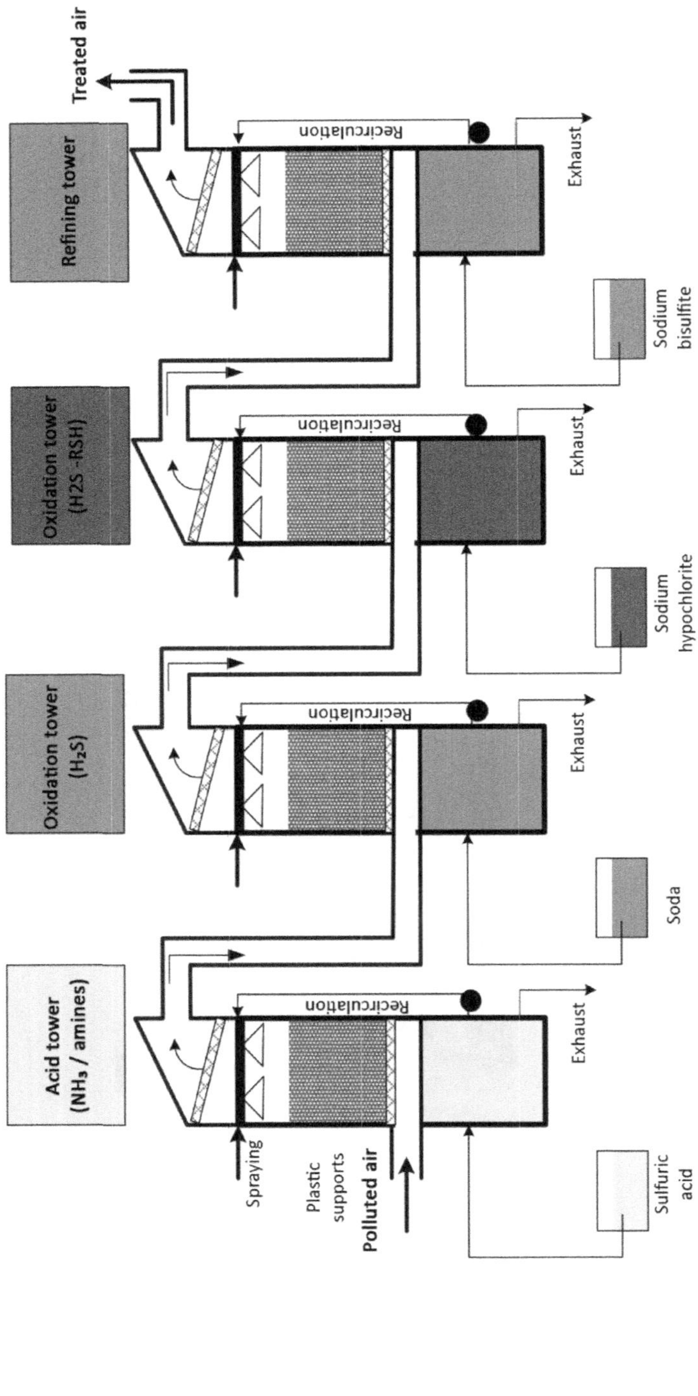

Figure 5.7. *Chemical odor treatment with four scrubber towers. For a color version of this figure, see www.iste.co.uk/gaid/wastewaterengineering4.zip*

5.3.1.4. *Safety recommendations*

The storage of reagents requires special attention and mandatory safety conditions:

– containment trays compatible with each type of reagent, with a volume conforming to current safety regulations;

– specific display screens with the appropriate security code displayed on the external walls of the reagent storage tanks or on the doors providing access to the storage areas;

– vents from the various reagent storage tanks to prevent a highly corrosive atmosphere in the room;

– tanks emptied via a pipe (open channels are prohibited);

– provision of a freeze-protected emergency shower and eyewash station is mandatory.

5.3.1.5. *Monitoring*

Odor control supervision is provided by:

– recording the operating time of the recycling pump;

– measurement of recirculation pressure;

– measurement of purge time;

– measurement of softened water volume;

– measurement of pH in the recirculation circuit (downstream of the recirculation pump for mixing optimization);

– in the case of oxidant-based purifiers, measurement of free chlorine concentration in the recirculation circuit;

– detection of very low, low and high levels in the scrubber;

– an audible alarm for a high level in the reagent tank;

– a reagent demand signal when the reagent tank level is low;

– stopping reagent injection when the reagent tank level is very low;

– detection of high and low levels in the brine preparation tank if the unit is equipped with an electrochlorinator.

5.3.2. *Biological treatment*

Based on the knowledge and experience gained with biological filters and biofilters, biological processes have been developed to eliminate gaseous pollutants.

5.3.2.1. *Heterotrophic biodeodorization*

The first type of biofilter to be implemented was peat biofiltration, capable of removing up to 99% of hydrogen sulfide, using a nutrient that is inexpensive compared to the large quantities of chemicals used in chemical deodorization treatments.

The process involves filtering the rising air introduced into the unit through a single duct connected to the plenum chamber at the base of the unit, below the suspended media bed. The first layer consists of gravel, which facilitates air distribution through the bed above. The biofilter is intermittently irrigated with water initiated according to a pre-set timer via a single nozzle. Its placement not only provides adequate irrigation but also allows easy access for maintenance. This monitored and controlled irrigation serves two functions: supplying the bacteria with water and nutrients (nitrogen and phosphorus) as needed and enabling the transfer of pollutants from the gaseous to the liquid phase.

The system's equilibrium is satisfactorily maintained by adding nutrients to the irrigation solution. Upstream gas injection can also be performed. This significantly improves the system's efficiency. The growth phase allows, for example, the transition from 10 to 10^{10} bacteria per gram of dry peat. Air is introduced and distributed in the lower section beneath a grid on which the absorbent material is placed. Regular top spraying ensures the medium is moistened and provides the necessary nutrients during the initial growth phase and throughout subsequent growth. The purification achieved in the treatment of gases from wastewater treatment plants is between 95% and 99% efficiency for sulfur compounds and amines, and 90% and 95% for ammonia, depending on the actual pollutant concentrations. Reactors operating with contact times of only a few seconds at gas loading rates of 1,000 $m^3/m^2/h$ under carefully controlled operating conditions are now well established.

However, the permissible air flow rate (approximately 100 $m.h^{-1}$) and the polluting load (< 10 g $S.m^{-3}.h^{-1}$) are relatively low, and significant maintenance is required to prevent pipe failure. Furthermore, the peat must be replaced after a few years of operation. These considerations make conventional biofiltration unsuitable for large installations.

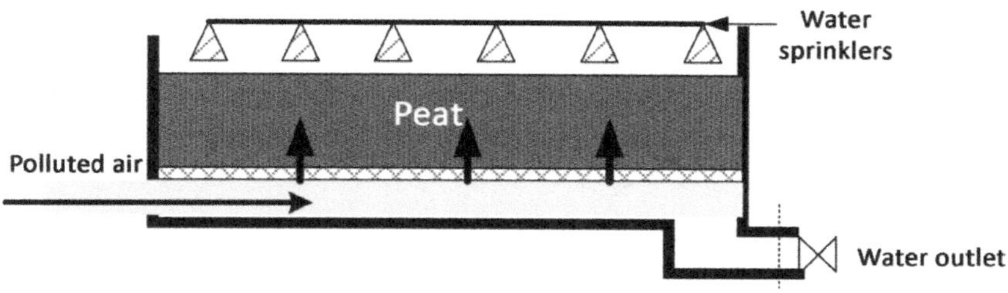

Figure 5.8. *Biofilter with organic support (peat)*

Figure 5.9. *Peat bed*

5.3.2.2. *Autotrophic biodeodorization*

A new process, combining the advantages of chemical treatment (compactness) and biological filtration (low consumption of chemicals or nutrients), has been developed. The autotrophic biofiltration system, such as the Alizair® (Veolia) process, is a technique that uses an air-circulating bioreactor lined with a medium for the growth of purifying bacteria. A mineral medium (volcanic rock, expanded clay, pumice, etc.) is used, promoting the growth of autotrophic, chemolithotrophic bacteria that oxidize foul-smelling organic matter. Furthermore, the long-lasting support is inert and does not become compacted as with peat.

5.3.2.2.1. Bacteria involved in autotrophic biodeodorization

Biological odor treatment involves microorganisms that oxidize malodorous organic and inorganic substances until odorless inorganic compounds are produced (CO_2, SO_4^{2-}, etc.).

In the context of biofiltration, the term is generally used to describe a curative treatment approach rather than a preventive one. This biological process relies on the metabolic transformations of gaseous pollutants through the incorporation of the harmful element into biomass and its transformation into a new chemical species (metabolites) that is not or only slightly harmful.

The support can play several roles:

– wetting and transfer of the pollutant;

– biomass fixation and retention;

– binding and retention of metabolites and enzymes;

– supply of trace elements or nutrients.

The substrate may be inert or non-inert. Gas–solid transfer occurs via the water used to moisten the organic substrate (peat, for example) or inert substrate (pozzolana, expanded clay, etc.). Moistened peat or compost is lightened by the addition of lighter elements, preventing compaction of the medium (branches, bark). The media provides the essential components required for biodegradation. Early applications of biofiltration relied on this principle and were primarily used to treat air with low pollutant concentrations (from a few milligrams to a few tens of $mg.Nm^{-3}$). Microorganisms attach to the substrate and break down the absorbed or adsorbed pollutants. This process is applicable to pollutants with low concentrations and air flow rates of approximately 100 $m.h^{-1}$.

Biodeodorization on an inert mineral support (autotrophic biodeodorization) relies on the degradation of sulfur pollutants by autotrophic bacteria with respect to the two main sources of matter necessary for bacterial metabolism, namely carbon (HCO_3^- and CO_2) and nitrogen (NH_4^+).

These chemolithotrophic bacteria use the energy released during the oxidation of sulfur to sulfate, an oxidation they themselves carry out. These are the bacteria of interest to us and they are most commonly used for odor elimination (when the odors are primarily composed of sulfur compounds). The most frequently encountered species are sulfate-oxidizing bacteria, sulfate-reducing bacteria and photosynthetic bacteria.

5.3.2.2.2. Bacteria involved in autotrophic biodeodorization

Autotrophic bacteria

Among the autotrophic bacteria are:

– Phototrophic bacteria, represented by algae and plants, as well as certain bacteria that use light energy to synthesize their organic matter from CO_2. The hydrogen donor is H_2O for plants and H_2S for sulfur bacteria, with the formation of S^{2-} in this specific case.

– Chemotrophic bacteria, which are represented by bacteria that oxidize various inorganic substances and use the energy released to synthesize organic matter from CO_2, which is their carbon source.

Sulfur-oxidizing bacteria

There are many types of sulfur-oxidizing bacteria, differing in both morphology and physiology.

The oxidation of H_2S may or may not involve the intermediate formation of elemental sulfur. The majority of the reactions observed during the deodorization process are carried out by non-photosynthetic bacteria, including:

– filamentous bacteria that accumulate intracellular sulfur globules: *Beggiatoa*, *Thiothrix* and *Thioploca*;

– non-filamentous bacteria that accumulate intracellular sulfur globules: *Achromatium*, *Thiovulum* and *Macromonas*;

– non-filamentous bacteria that do not accumulate intracellular sulfur globules: *Thiobacillus*, *Thiomicrospira*, *Thiodendron* and *Sulfolobus*.

The reaction mechanism is as follows:

– aerobic oxidation of sulfides to elemental sulfur (with possible accumulation of sulfur in the cytoplasm in the form of granules):

$$H_2S + 1/2\ O_2 \rightarrow SO + H_2O$$

– oxidation of elemental sulfur into sulfates:

$$SO + 3/2\ O_2 + 3\ H_2O \rightarrow SO_4^{2-} + 2\ H_3O^+$$

These bacteria use CO_2 as their sole carbon source.

Thiobacillus are responsible for oxidation into sulfates, sulfides, thiosulfates, elemental sulfur and thiocyanates.

Depending on whether the environment is aerobic or anaerobic, the reaction are as follows:

– In an aerobic environment:

- with *Thiobacillus thiooxidans*:

$$SO + 3/2\ O_2 + 3\ H_2O \rightarrow SO_4^{2-} + 2\ H_3O^+$$

- with *Thiobacillus thioparus*:

$$H_2S + 2\ O_2 + 2\ H_2O \rightarrow SO_4^{2-} + 2\ H_3O^+$$

$$S_2O_3^{2-} + 2\ O_2 + 3\ H_2O \rightarrow SO_4^{2-} + 2\ H_3O^+$$

- with *Thiobacillus thiocyano-oxidans*:

$$SCN^- + 2\ O_2 + 2\ H_2O \rightarrow SO_4^{2-} + NH_4^+ + CO_2$$

These *Thiobacillus* are Gram-negative, spore-forming, bacillary-shaped bacteria, sized between 1 and 3 μ. *T. thioparus* develop at a pH close to neutral, and their growth ceases approximately pH 5. *T. thiooxidans* only develop at pH < 5, and their growth can lower the pH to 0.

– In an anaerobic environment, facultative anaerobic *Thiobacillus*, such as *Thiobacillus denitrificans*, use nitrates as an oxygen source:

$$5\ SO + 6\ NO_3^- + 6\ H_2O \rightarrow 3\ N_2 + 5\ SO_4^{2-} + 4\ H_3O^+$$

$$6\ S_2O_3^{2-} + 10\ NO_3^- + 3\ H_2O \rightarrow 5\ N_2 + 12\ SO_4^{2-} + 2\ H_3O^+$$

$$0.418.\ (H_2S + HS^-) + 1.5\ O_2 + 0.078\ (HCO_3^- + NH_4^+) + 0.312\ CO_2 \rightarrow$$

$$0.836\ SO_4^{2-} + 0.078\ C5H7NO_2 + 0.382\ H^+ + 0.358\ H_2O$$

The optimum pH for the growth of these bacteria is between 6.2 and 7.4.

Sulfate-reducing bacteria

The most commonly encountered genus is *Desulfovibrio desulfuricans*. These are non-spore-forming, Gram-negative, highly motile bacteria, 1–2 μm in size and strict anaerobes. They reduce sulfates to sulfides, most often using organic compounds such as dicarboxylic acids, mercaptans, pyruvates, etc., as an energy source.

In the presence of pyruvate, we obtain:

$$1/4\ SO_4^{2-} + CH_3COCOOH \rightarrow 1/4\ S^{2-} + CH_3COOH + CO_2$$

In the presence of methyl mercaptan, we obtain:

$$SO_4^{2-} + CH_3SH \rightarrow 2\ S^{2-} + CO_2 + 2\ H_2O$$

Many authors place this type of bacteria between autotrophs and heterotrophs because they do not use CO_2 as with autotrophs, they derive their energy from the anaerobic oxidation of hydrogen. Table 5.10 summarizes the different activities of autotrophic microorganisms involved in the oxidation of sulfur compounds.

Micro organisms	Type	Use of organic compounds	Sulfur compounds used	pH
Autotrophs	Aerobic Facultative aerobic	–	H_2S, SOx, $S_2O_3^{2-}$	0.6–10
Facultative autotrophs	Aerobic	+/–	H_2S, SOx, $S_2O_3^{2-}$	0.6–10
Phototrophs	Anaerobic	+/–	H_2S, SOx, $S_2O_3^{2-}$	4.5–10

Table 5.10. *Activities of autotrophic microorganisms*

5.3.2.2.3. Mode of action of autotrophic bacteria

Biological odor treatment involves microorganisms that oxidize malodorous organic and inorganic substances until odorless inorganic compounds are produced (CO_2, SO_4^{2-}, etc.). As in biological wastewater treatment, the organic compounds present in gaseous effluents provide substrates that support both the catabolic (energy-producing) and anabolic (biomass-producing) pathways of bacterial metabolism. Autotrophic bacteria will eliminate inorganic substances such as hydrogen sulfide and sulfites, which will be oxidized to sulfates according to the reactions:

$$H_2S + 1/2\ O_2 \rightarrow SO + H_2O$$

$$SO + 3/2\ O_2 + 3\ H_2O \rightarrow SO_4^{2-} + 2\ H_3O^+$$

$$H_2S + 2\ O_2 + 2\ H_2O \rightarrow SO_4^{2-} + 2\ H_3O^+$$

$$S_2O_3^{2-} + 2\ O_2 + 3\ H_2O \rightarrow SO_4^{2-} + 2\ H_3O^+$$

In parallel, bacterial synthesis will be carried out according to the reaction:

$$H_2S + NH_3 + O_2 + HCO_3^- + PO_4^{3-} \rightarrow C_5H_7NO_2\ (S, P) + H_2O + SO_4^{2-}$$

These symbolic equations represent the sulfur transformation pathways. The two possible pathways are as follows:

– assimilation by microorganisms ($C_5H_7NO_2$);

– transformation into non-polluting or low-polluting products (CO_2, SO_4^{2-}, etc.).

Thus, all sulfur compounds will be oxidized to sulfates (and possibly sulfites).

5.3.2.2.4. Reactor operation

While biological wastewater treatment uses both suspended-growth and attached-growth reactors, biodeodorization systems primarily rely on attached-growth processes. In these systems, microorganisms develop on a mineral or organic support medium such as peat, soil, gravel, expanded clay, activated carbon or synthetic materials (e.g. polypropylene or polyethylene). The support medium allows:

– the attachment of purifying bacteria;

– the absorption and/or adsorption of organic compounds to be eliminated.

It has been observed that the flow of the liquid film around the grain is better with an inert mineral support (such as Biodagene, for example) than with an organic material such as peat. Adsorption and diffusion reactions are improved. The principle of biological deodorization involves circulating the gaseous effluent to be treated through a reactor containing a support material on which bacteria have grown. To optimize the biodeodorization process, it is essential that the materials to be removed be water-soluble. This requires constant humidification of the bed to ensure that the transfer of the materials to be removed by the bacteria occurs under the best possible conditions. Inadequate humidification leads to drying of the substrate and a decrease in bacterial activity. This is prevented by circulating irrigation water in a counter-current flow within the reactor. During the first few weeks of reactor operation, the irrigation water is continuously recirculated to seed the reactor. It is also important to ensure the supply of nutrients (N, P) if the medium is deficient. This N and/or P supply is achieved by introducing the necessary elements in the irrigation water. Oxygen is supplied by the gaseous effluent that passes through the reactor.

The time it takes for the gaseous effluent to pass through the reactor can vary from 3 to 8 seconds. It depends on the concentration of the substances to be removed, the flow rate to be treated and the type of packing.

Most odor-producing compounds are fairly water-soluble. However, some compounds, such as mercaptans (RSH), are only slightly water-soluble. This creates a kind of chemical reaction + bioremediation association.

Autotrophic biodeodorization requires a source of mineral carbon. This can be:

– provided in the form of CO_2 upstream in the gaseous effluent;

– supplied by the carbonates present in the irrigation water;

– provided by the support material (maerl, for example), which also serves as a support for absorption and fixation of microorganisms.

Biological deodorization on fixed beds can be carried out:

– on an inert mineral substrate (autotrophic biodeodorization) using autotrophic bacteria;

– on an organic substrate (heterotrophic biodeodorization) using heterotrophic bacteria.

With an adequate supply of nutrients, autotrophic biodeodorization leads to H_2S and nitrogen compound removal efficiencies of 99% for an air loading rate of 500 $m.h^{-1}$.

Material	**Mineral types: biodagene, pozzolana, expanded shale** **Granulometry = 3–6 mm**
Reactor	Concrete pit protected by an epoxy coating (with respect to acidic pH) Reaction time < 20 s Air loading rate = 300–500 $m^3.m^{-2}.h^{-1}$ Height = 1–2 m
Bacteria	*Thiobacillus*
Gaseous effluent	Flow rate = 2,500 $Nm^3.h^{-1}$
Sprinkling	Achieved using nozzles positioned above the filter bed Instantaneous flow rate: 10–20 $l.m^{-3}$ $material.h^{-1}$ Sprinkler frequency: 2 minutes every 30 minutes

Table 5.11. *Operating conditions of a biological deodorization process*

The filter consists of a concrete pit protected by a coating that limits concrete degradation (epoxy coating). The coating must protect the concrete from aggressive pH conditions (acidic pH). For small treatment units (< 5,000 $Nm^3.h^{-1}$), it is possible to use polyethylene tanks. A floor provides support for the material and ensures good air distribution within the biofilter. This floor consists of gratings (20–30) covered with plastic mesh (4 x 4). A layer of gravel (6–10 mm) is placed on top of the wire mesh to a height of 30 cm to promote air circulation, followed by the support (3–6 mm) to a height of 1–2 m. At an air loading rate of 500 $m.h^{-1}$, due to significant bacterial growth, the pressure drop is 50 mm of water when there are high H_2S concentrations (15 $mg.Nm^{-3}$ air) and 25 mm of water when there are low H_2S concentrations (< 15 $mg.Nm^{-3}$ air). The foundation slab is designed with a 2% slope to collect the sprayed water. The siphons for draining this water are designed for a bed head loss of approximately 90 mm CE. Except during the seeding period, this water is returned to the head of the treatment plant or to the biological treatment process. Covering the filter is recommended to prevent it from drying out too quickly during periods of high heat, which can cause malfunctions in the purification process. Similarly, during periods of heavy rain, covering the filter prevents the biomass from being washed away.

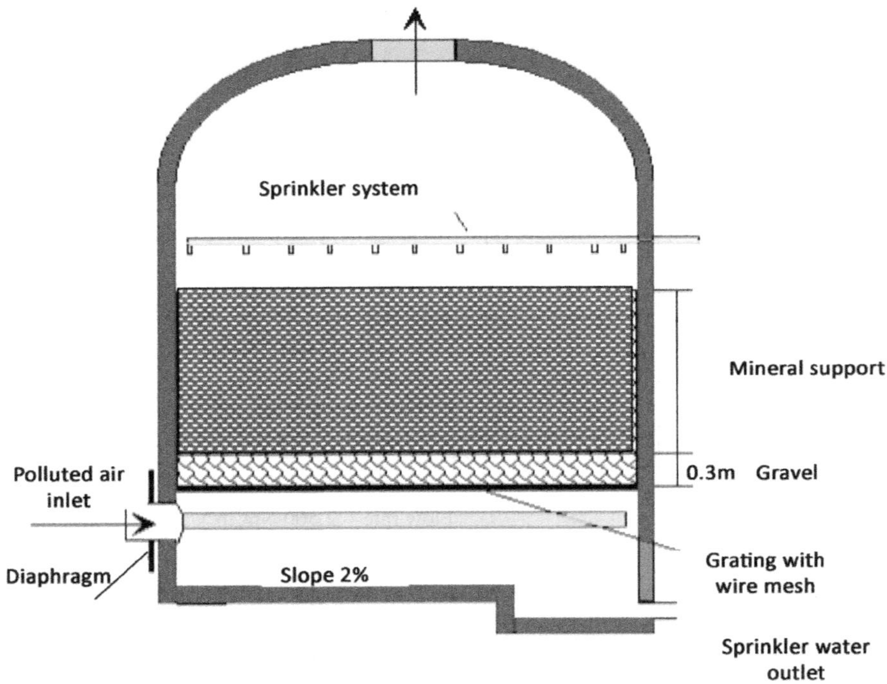

Figure 5.10. *Cross-sectional diagram of an autotrophic biofilter*

The biofilter must be thoroughly watered throughout its entire volume. Watering is key to successful commissioning. Watering serves to moisten the filter and supply nutrients and normally involves approximately 0.03 $l.Nm^{-3}$ of air treated (h^{-1}). Watering is carried out in bursts of 1 minute 30 second every 15 minutes. The sprinkler heads must be positioned so that watering is uniform across the entire bed surface. Regardless of the filter's operating conditions, watering can be supplied from:

– groundwater;

– dechlorinated mains water;

– treated wastewater (this must not contain easily assimilable organic matter, at the risk of developing heterotrophic biomass, which could become predominant over autotrophic biomass).

The bed is sprayed by nozzles whose number and arrangement are such that the spraying is uniform over the surface of the bed. Sprayers must have an internal geometry free of dead zones, an orifice that is neither too small nor too narrow, and a minimum number of bends to prevent clogging by biomass during spray water recycling or by any salts (such as calcium phosphate).

The polluted air intake must be designed so that the gas is distributed evenly over the filter. The cross-sections required are sized to achieve a maximum velocity of 8 $m.s^{-1}$. The installation is easier to adjust thanks to diaphragms.

The nutritional requirements of autotrophic biomass are as follows:

– a source of mineral carbon (CO_2 or carbonates);

– a source of nitrogen and phosphorus:

- trisodium phosphate;
- ammonium phosphate;
- ammonium nitrate;
- NH_3.

Nutrients can either be initially present in sufficient quantities in the gas and spray water or be introduced into the gas phase or liquid phase by controlled injection.

There are no temperature limits for polluted air. Classical values of 10°C to 40°C are temperatures that do not disrupt the biofilter efficiency. Care is taken during cold periods (air < 5°C), for example, by heating the air, if necessary, to prevent the biofilter from freezing.

This operating mode, with recycling of irrigation water, is implemented during the initial stages of the biological process to promote bacterial seeding and growth. In this case, a tank (volume = 150 L) to collect the water from the biofilter outlet must be installed. This tank also serves to regulate the spray water to the optimum pH, as the water returning to the tank is acidic.

The regulating solution is basic: it is advisable to use KOH and avoid NaOH, which could lead to excessively high sodium concentrations in the spray water, harmful to bacterial activity.

The irrigation solution is drained once the biofilter is operating at a stable level and is well seeded. The irrigation solution is then discharged directly into the sewer or back into the head of the treatment plant. The tank initially used during the start-up phase is then used as a mixing tank for the nutrient solutions.

5.3.2.2.5. Refining with granular activated carbon

If it is necessary to comply with the odor thresholds for mercaptans and organic compounds (aldehydes, ketones), further treatment may be required by filtration through granular activated carbon, placed downstream of the biofilter, or a chemical deodorization tower operating at pH 11 and in the presence of bleach (elimination of RSH).

Note that RSH concentrations are low (< 2 mg/Nm-3) when treating sludge from prolonged aeration or low loading. In other cases (fresh sludge, for example) or when there is confinement, RSH concentrations can be higher and therefore require additional treatment.

Under optimal operating conditions, the biofilter's purification efficiency with respect to methyl mercaptan (CH3SH) varies between 50 and 90% and does not allow the nuisance threshold (0.07 $mg.m^{-3}$) of this pollutant in the treated air to be reached.

It is therefore necessary to implement an air purification system using activated carbon to protect against odors related to methyl mercaptan, particularly when dwellings are nearby. A residence time of a few seconds and a velocity of 0.2 $m.s^{-1}$ (720 $m.h^{-1}$) are optimal as a treatment option.

Given the concentrations downstream of the biofilter, the cycle time between two GAC regenerations is between 18 and 24 months. The activated carbon used must be designed for a "deodorization" application, meaning it must be capable of adsorbing sulfur molecules, traces of amines and, if possible, organic molecules such as aldehydes and ketones.

Since the spray water may be carried by the air and humidify the GAC, a demister-type device can be placed between the biofilter and the carbon filter or even within a separate unit.

When the activated carbon is saturated, it can be regenerated on site chemically with caustic soda when possible.

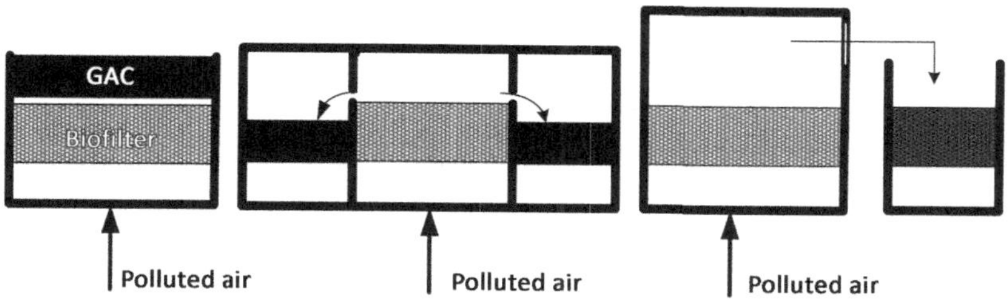

Figure 5.11. *Different configurations of the activated carbon filter*

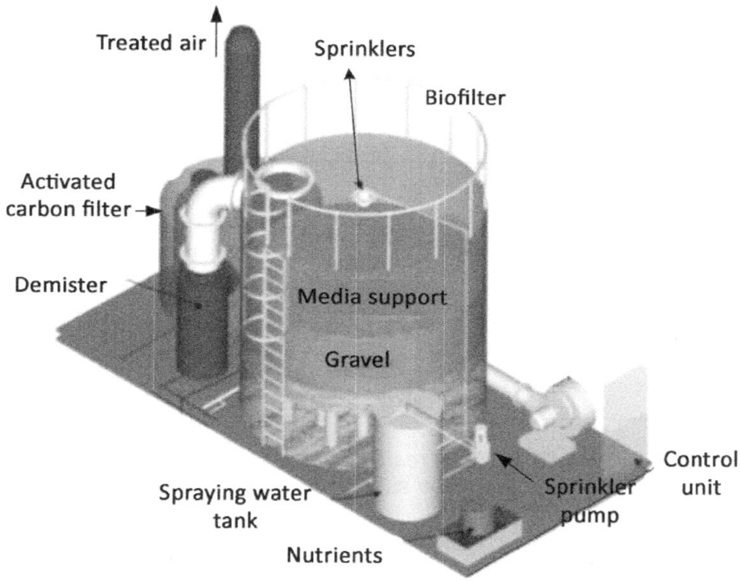

Figure 5.12. *Granular activated carbon*

5.3.2.2.6. Performance

Autotrophic biodeodorization involves small structures and requires little maintenance. It is particularly recommended when the concentration of mercaptans in the gaseous effluents is ≤ 1 mg.Nm^{-3} (efficiency > 99%). Mercaptan removal efficiencies are approximately 90% when their inlet concentration reaches 2 mg.Nm^{-3}. When the concentration of mercaptans (RSH) is higher (> 2 mg.Nm^{-3}), the mercaptan purification efficiencies obtained at the biofilter outlet are therefore approximately 70%.

Conditions	Nutrients present in the irrigation water	Additions	H_2S efficiency	Estimate of reagent consumption
Recycling spray water and regulation at pH = 7	NH_4^+:10–300 mg N.L^{-1} PO_4^{2-}: 0–5 mg P.L^{-1} Alkalinity < 100 mg.L^{-1} $CaCO_3$	KOH for regulation	Rdt H_2S > 99.5% if H_2S < 40 mg.m^{-3}	Depends on the H_2S concentration at the inlet
Drainage of spray water Without recirculation	NH_4^+ < 0.2 mg N.L^{-1} PO_4^{2-} < 0.03 mg P.L^{-1} Alkalinity < 300 mg.L^{-1} $CaCO_3$	–	Rdt H_2S > 99.8% if H_2S < 15 mg.m^{-3} Rdt H_2S > 99.5% if H_2S < 50 mg.m^{-3} Rdt H_2S > 98% if H_2S < 100 mg.m^{-3}	–
Disposal of spray water Without recirculation Enrichment with external input of N and P	NH4$^+$ < 0.2 mg N.L^{-1} PO_4^{2-} < 0.03 mg P.L^{-1} Alkalinity < 300 mg.L^{-1} $CaCO_3$	NH_4Cl $(NH_4)_2HPO_4$ KH_2PO_4 NH_4: 10–15 mgN.L^{-1} PO_4^{2-}: 3–5 mg P.L^{-1}	Rdt H_2S > 99.9% if H_2S < 50 mg.m^{-3} Rdt H_2S > 99.8% if H_2S <100 mg.m^{-3}	Consumption per 1,000 m^3 of treated air: NH_4Cl: 1.52 g; KH_2PO_4: 0.87 g

Table 5.12. *Main results of biodeodorization on an inert support (France)*

For NH_3 concentrations less than 5 mg.Nm^{-3}, the biofilter demonstrates its strong capabilities (efficiency greater than 95%).

NH_3 inlet	NH_3 outlet	Yield
< 5 mg.Nm^{-3}	< 0.5 mg.Nm^{-3}	> 90%

Table 5.13. *Reduction of NH_3 on the inert support biofilter*

The results obtained show that methyl mercaptan is eliminated with efficiencies of:

– 99% when its input concentration is ≤ 1 mg.Nm^{-3};

– 90% when its input concentration is > 1 mg/Nm^3 and ≤ 2 $mg.Nm^{-3}$;

– 70% when its input concentration is > 2 $mg.Nm^{-3}$.

At high concentrations, CH_3SH removal is difficult for several reasons:

– CH_3SH is difficult to dissociate.

– CH_3SH partially oxidizes into sulfonic acid (a poorly biodegradable intermediate product).

– CH_3SH has low solubility.

Upon initial setup, the biofilter reaches its optimal capacity after approximately one month. A prolonged interruption of the system (water or power outage lasting from a few days to a few weeks) has little impact on biological activity, and purification returns to optimal levels within 3–10 days. Restarting is even faster (one to two days maximum) when the inlet gas concentrations are low.

5.3.2.2.7. Dimensioning principles

Determining the size of an autotrophic biodeodorization plant should be based on the following data: flow rate for treatment (Q, $m^3.h^{-1}$); concentrations of gaseous pollutants to be treated (H_2S, RSH (possibly R_2SH), NH_3, organic N); and use of industrial water as spray water.

The biofilter sizing is determined as follows:

– Polluted air velocity in the biofilter: v = 300–500 $m.h^{-1}$.

– Packed bed depth = 1–2 m.

– Gravel depth: H = 0.3 m.

– Average spraying water flow rate: Qspw = 0.03 $l.m^3$ $air.h^{-1}$.

– Spraying water frequency and duration: tspw = 1.5 min every 15 min.

– Total area (m^2) = polluted air flow rate/superficial velocity, or for V = 500 $m.h^{-1}$: S = air flow rate/500.

– Volume of bacteria-fixing material/unit: V_B (m^3) = S x H.

– Volume of gravel/unit: V_G (m^3) = S x H_G.

– Contact time (sec) = (V_B x 3 600)/Q air.

– Number of nozzles per area:

Nb = S/(d sin 45)^2 for filters with a rectangular or square cross-section

Nb = D^2/(d sin 45)^2 for filters with a circular cross-section

– Distance between two sprinklers (mm) = d.sin45.

– Number of sprinklers = L/d.sin45.

– Nozzle center distance = d.sin45.

– Spraying water flow rate per spraying system and per unit: Q bar (l/mn) = Nb x Q nozzle.

– Spraying water velocity: V = 2 m.s^{-1}.

– Spraying system diameter:

$$D_{spraying\ system}\ (mm) = 2.10^3 \sqrt{\frac{D_{spraying\ system}}{1000 \times 60 \times \pi \times V'}}$$

EXAMPLE.– Case of low RSH concentration

– Polluted air flow rate to be processed = 8,500 m^3.h^{-1}.

– Concentrations of gaseous pollutants to be treated:

H_2S = 4.8 mg.Nm^{-3}; RSH = 0.9 mg.Nm^{-3}; NH_3 = 1.2 mg.Nm^{-3}; organic N = 1.2 mg.Nm^{-3}

– Polluted air velocity in the biofilter = 500 m.h^{-1}.

– Total material depth = 1.3 m, distributed as follows:

- height of bacteria-fixing material (inert support) = 1 m;
- gravel depth = 0.3 m.

– Spraying water frequency and duration = 1.5 min every 15 min.

– Number of nozzles = 1.5 nozzles.m^{-2}.

– Total surface area (m^2) = 8,500/500 = 17.

– Number of units = 17/10 = 1.7.

– Number of units retained (rounded) = 2.

– Total surface area required (m^2) = 2 x 10 = 20.

– Volume of bacteria-fixing material/unit (m^3) = 10 x 1 = 10.

– Total volume of bacteria-fixing material (m^3) = 2 x 10 = 20.

– Volume of gravel/unit: V'_G (m^3) = 10 x 0.3 = 3.

– Total volume of gravel: VG = 2 x 3 = 6.

– Contact time (sec) = (20 x 3,600)/8,500 = 8.47 s.

– Spraying water flow rate: Qspw ($l.h^{-1}$) = 0.03 x 8,500 = 255.

– Instantaneous spraying water flow rate: Q inst ($m^3.h^{-1}$) = 255 x 10^{-2} = 2.55 $m^3.h^{-1}$.

– Number of nozzles retained per unit area: Nbu = 10/(0.800 x 0.70) ^{2} = 31.88 for filters with rectangular or square cross-section and with solid cone nozzles without divergence.

– Number of nozzles retained per unit area = 32 (rounded).

– Total number of nozzles = 32 x 2 = 64.

– Nozzle center distance = 0.800 x 0.700 = 0.56 m.

– Spraying water flow rate/nozzle and per unit: Q nozzle ($l.mn^{-1}$) = (2.55 x 1,000)/(32 x 60) = 1.32.

– Total spraying water flow rate required: Q′ ($l.mn^{-1}$) = 1.32 x 32 x 2 = 84.48.

– Number of sprinklers = 4/(0.800 x 0.700) = 7.14 #7.

– Number of nozzles per spraying system = 2.5/(0.800 x 0.700) = 4.46 (assuming a filter width of 2.5 m).

– Total number of nozzles/unit = 7.14 x 4.46 = 31.84 = 32.

The global lines of the spraying system are arranged with alternating sections consisting of a spraying system with four nozzles and a spraying system with five nozzles.

– Flow rate per spraying system and per unit: Q spraying system ($l.mn^{-1}$) = 4 x 1.32 = 5.28 and 5 x 1.32 = 6.6.

– Diameter of a spraying system:

$$\text{Dspraying system (mm)} = 2{,}000\sqrt{\frac{6.60}{1{,}000 \text{x} 60 \text{x} 2\pi}} = 8.36 \text{ for } V' = 2 \text{ m.s}^{-1}.$$

– Diameter (mm) = 8.40–8.50 (rounded).

– Number of distribution openings/unit: Opening = 10/10 = 1.

– Number of openings: N = 1.

– Total number of openings required: N = 2 x 1 = 2.

– Polluted air flow rate/opening ($m^3.h^{-1}$) = 8,500/2 = 4,250.

– Cross-section: S (m²) = 4,250/3,600 x 8 = 0.15 where V″ = 8 $m.s^{-1}$.

– Diameter: D opening (mm) $= 2{,}000\sqrt{\frac{0.15}{\pi}} = 437.$

– Removal of H_2S > 99.8%, that is, H_2S < 0.1 $mg.Nm^{-3}$ air.

– Removal of NH_3 > 90%, that is, NH_3 < 0.12 $mg.Nm^{-3}$ air.

– Removal of amines > 90% that is, amines < 0.12 $mg.Nm^{-3}$ air.

– Removal of RSH > 99%, that is, RSH < 0.07 $mg.Nm^{-3}$ air.

EXAMPLE.– Case of a high concentration of RSH.

– Polluted air flow rate to be processed = 8,500 $m^3.h^{-1}$.

– Concentrations of gaseous pollutants to be treated: H_2S = 4.8 $mg.Nm^{-3}$; RSH = 3.2 mg.Nm-3; NH_3 = 1.2 $mg.Nm^{-3}$; organic N = 1.2 $mg.Nm^{-3}$.

– Total surface area (m^2) = 8,500/500 = 17.

– Number of units = 17/10 = 1.7.

– Number of units retained (rounded) = 2.

– Total surface area required (m^2) = 2 x 10 = 20.

– Volume of bacteria-fixing material/unit (m^3) = 10 x 1 = 10.

– Total volume of bacteria-fixing material = 2 x 10 = 20.

– Volume of gravel/unit: V'_G (m^3) = 10 x 0.3 = 3.

– Total gravel volume: V_G = 2 x 3 = 6.

– Contact time (sec) = (20 x 3,600)/8,500 = 8.47 s.

– Spraying water flow rate: Qspw ($l.h^{-1}$) = 0.03 x 8,500 = 255.

– Instantaneous spraying system flow rate: Q inst ($m^3.h^{-1}$) = 255 x 10^{-2} = 2.55 $m^3.h^{-1}$.

– Number of nozzles retained per unit area: Nbu = 10/(0.800 x 0.7)2 = 31.88 for filters with rectangular or square cross-sections and with solid cone nozzles without divergence.

– Number of nozzles retained per unit area = 32 (rounded).

– Total number of nozzles = 32 x 2 = 64.

– Nozzle center distance = 0.800 x 0.700 = 0.56 m.

– Spraying water flow rate per nozzle and per unit: Q nozzle ($l.mn^{-1}$) = (2.55 x 1,000)/(32 x 60) = 1.32.

– Total spraying water flow rate required: Q′ ($l.mn^{-1}$) = 1.32 x 32 x 2 = 84.48.

– Number of sprinklers = 4/(0.800 x 0.700) = 7.14 #7.

– Number of nozzles per spray bar = 2.5/(0.800 x 0.700) = 4.46 (assuming a filter width of 2.5 m).

– Total number of nozzles/unit = 7.14 x 4.46 = 31.84 #32.

The lines are arranged with alternating sections consisting of spray bars with four nozzles and spray bars with five nozzles.

– Flow rate per spraying system and per unit: Q spraying system ($l.mn^{-1}$) = 4 x 1.32 = 5.28 and 5 x 1.32 = 6.60.

– Diameter of a sprinkler: $D\ bar\ (mm) = 2{,}000\sqrt{\frac{6.60}{1{,}000x60x2\pi}} = 8.36$ for $V' = 2\ m.s^{-1}$).

– Diameter (mm) = 8.40–8.50 (rounded).

– Number of distribution openings/unit: Opening = 10/10 = 1.

– Number of openings = 1.

– Total number of openings required: N = 2 x 1 = 2.

– Polluted air flow rate/opening ($m^3.h^{-1}$) = 8,500/2 = 4,250.

– Cross-section: S (m^2) = 4,250/3,600 x 8 = 0.15, where $V'' = 8\ m.s^{-1}$.

– Diameter: $D\ opening\ (mm) = 2{,}000\sqrt{\frac{0.15}{\pi}} = 437.$

– Removal of H_2S > 99.8%, that is, H_2S < 0.1 $mg.Nm^{-3}$ air.

– Removal of NH_3 > 90%, that is, NH_3 < 0.12 $mg.Nm^{-3}$ air.

– Removal of amines > 90%, that is, amines < 0.12 $mg.Nm^{-3}$ air.

– Removal of RSH > 70%, that is, RSH < 0.96 $mg.Nm^{-3}$ air (exceeds the guaranteed thresholds).

An activated carbon filter must then be placed downstream of the biofilter.

Activated carbon filter

– Usable area: S GAC (m^2) = 8,500/(0.2 x 3,600) = 12 for velocity = 0.2 $m.s^{-1}$.

– Usable height: H GAC (m) = 0.2 x 3 = 0.6 m.

– Usable volume (m^3) = 12 x 0.6 = 7.2.

– Contact time (sec) = 3.

– H_2S outlet concentration ($mg.Nm^{-3}$) < 0.01.

– RSH outlet concentration ($mg.Nm^{-3}$) < 0.07.

The intake should be within the following concentrations:

– Phosphorus: the PO_4^{3-} concentration must be between 3 and 5 mg $P.L^{-1}$ of sprinkler water, that is, 160 mg P/1,000 m^3 of treated air.

– Nitrogen: the NH_4^+ concentration must be between 10 and 15 mg $N.L^{-1}$ of sprinkler water, which corresponds to a minimum contribution of 400 mg N/1,000 m^3 of treated air. Nitrogen can be supplied as a nutrient, either through the recirculating water or directly into the air stream. If the NH_3 concentration in the gas phase is greater than 1 $mg.Nm^{-3}$, the addition of NH_4^+ is not required. However, if NH_3 (gas) < 1 $mg.Nm^{-3}$, nitrogen intake may be insufficient, and its addition is advised at the concentrations seen previously.

– Carbonates: the supply from the sprinkler water must be 7.5g CO_3^{2-}/1,000 m^3 of treated air.

Nitrates and nitrites contained in the irrigation water are eliminated in the biofilter.

Consumption of chemical reagents and activated carbon

The consumption of chemical reagents and activated carbon (when used) is estimated based on the standard air quality to be treated. Table 5.14 provides the optimal values (in kg/1,000 m^3 of air).

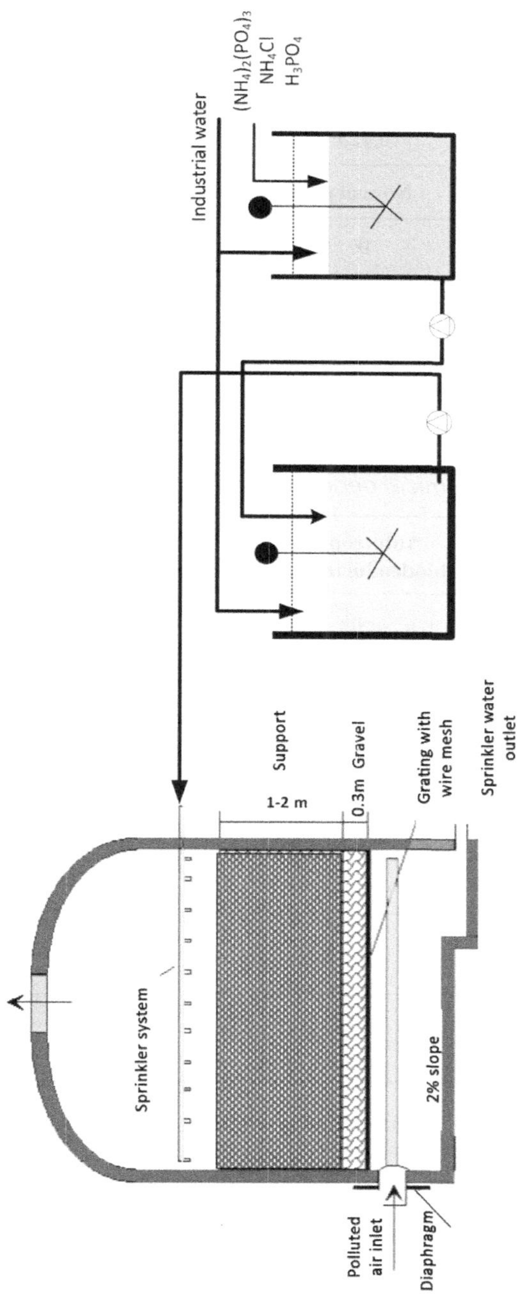

Figure 5.13. *Steady-state process diagram*

	Air with low pollution	Moderately polluted air	Very polluted air
Seeding phase			
KOH 45% (weight)	0.125	0.125	0.125
Normal operating phase			
H3PO4 75%	0	0	0.0005
Activated carbon	0.009	0.018	0.027
Water	40	40	40

Table 5.14. *Reagent consumption according to air pollution levels*

Comparison between chemical deodorization and biodeodorization

	Chemical deodorization	Autotrophic biodeodorization	Heterotrophic biodeodorization	GAC filter	Bioautotroph + GAC
H_2S	+++	+++	++	+++	+++
RSH	++	+	++	+++	+++
NH_3	+++	++	++	+++	+++
RNH_2	++	+	++	+++	+++
VOC	++	–	++	+++	+++
Odor unit	+++	–	+	+++	+++
Investment cost	+++	+	++	++	++
Operation cost	+++	–	+	+++	++

Table 5.15. *Performance and operation comparison of some deodorizing processes*

Table 5.16 summarizes the main sizing and operational parameters of the two types of biodeodorization.

a) b)

Figure 5.14. *Deodorization: a) chemical and b) biological*

Comparison between autotrophic and heterotrophic biodeodorization

Parameters	Autotrophic biodeodorization	Heterotrophic biodeodorization
Throughput	500 m.h^{-1}	100 m.h^{-1}
Pressure losses	0.05 m of water column	0.080–0.100 m of water column
H_2S and mercaptans removal	Very good 99% when [RSH] ≤ 1 mg.Nm^{-3} 90% when [RSH] > 1 and ≤ 2 mg/Nm^3 Approx. 70% when [RSH] > 2 mg.Nm^{-3}	Very good 99% for [RSH] ≤7 mg.Nm^{-3}
Ammonia	Very good	Very good
Amines	Very good	Very good
Preferential pathways	No	Often observed
Material compaction	No	Often observed
H_2S treatment load	50 g.m^{-3}.h^{-1}	10 g.m^{-3}.h^{-1}
Reaction to load variations	No	No
Nutrient addition	N, P, CO_3^{2-}	N, P, C organic (glucose, methanol)
Reaction to the prolonged shutdowns	Restart observed immediately or after one week of operation	Restart observed faster than autotrophic system
Additional treatment	Activated charcoal if concentration of inlet RSH > 2.5 mg.Nm^{-3}	No

Table 5.16. *Comparison between autotrophic and heterotrophic biodeodorization methods*

Compared to the heterotrophic process, we can note the following advantages of autotrophic biodeodorization:

– higher air loading rate;

– lower pressure losses;

– no preferential flow paths;

– no material compaction;

– superior processing capacity;

– responds quickly to load variations.

5.4. References

Alinezhad, E., Haghighi, M., Rahmani, F., Keshizadeh, H., Abdi, M., Naddafi, K. (2019). Technical and economic investigation of chemical scrubber and biofiltration in removal of H_2S and NH3 from wastewater treatment plant. *J. Environ. Manag.*, 241, 32–43.

Barbosa, L. and Stuetz, R.M. (2013). Performance of activated sludge diffusion for biological treatment of hydrogen sulphide gas emissions. *Water Sci. Technol.*, 68, 1932–1939.

Barbusinski, K., Kalemba, K., Kasperczyk, D., Urbaniec, K., Kozik, V. (2017). Biological methods for odor treatment – A review. *J. Clean. Prod.*, 152, 223–241.

Bolton, E., Brennan, B., Donlon, M. (1994). Peat biofiltration as an odour control technology. In *Symposium on Odour Control and Prevention in the Water Industry*, Newcastle upon Tyne.

Bonino, P. (2000). Treatment of hydrogen sulfide using the Lanodor process. In *1st World Water Congress*, Paris.

Bonnin, C., Darmon, D., Laplanche, A. (1994). Deodorization in wastewater treatments plants by wet-scrubbing on packed column and chlorine oxidation. In *Symposium on Odour Control and Prevention in the Water Industry*, Newcastle upon Tyne.

Burgess, J.E., Parsons, S.A., Stuetz, R.M. (2001). Developments in odour control and waste gas treatment biotechnology: A review. *Biotechnol. Adv.*, 19, 35–63.

Gaid, K., Cretenot, C., Sibony, J., Bonnin, C. (1995). Autotrophic biological deodorization of foul-smelling gases at STPs. *World Water and Environmental Engineering*, October, 28–29.

Kasperczyk, D., Urbaniec, K., Barbusinski, K., Rene, E.R., Colmenares-Quintero, R.F. (2019). Application of a compact trickle-bed bioreactor for the removal of odor and volatile organic compounds emitted from a wastewater treatment plant. *J. Environ. Manag.*, 236, 413–419.

Le Sauze, N., Laplanche, A., Martin, G., Paillard, H. (1991). A process of washing and ozonation to deodorize an atmosphere contaminated by sulfides. *J. Int. Ozone Assoc.*, 13(3), 331–347.

Lewkowska, P., Cieslik, B, Dymerski, T., Konieczka, P., Namiesnik, J. (2016). Characteristics of odors emitted from municipal wastewater treatment plant and methods for their identification and deodorization techniques. *Environ. Res.*, 151, 573–586.

Lin, S., Mackey, H.R., Hao, T., Guo, G., van Loosdrecht, M.C.M., Chen, G. (2018). Biological sulfur oxidation in wastewater treatment: A review of emerging opportunities. *Water Res.*, 143, 399–415.

Liu, J., Kang, X., Liu, X., Yue, P., Sun, J., Lu, C. (2020). Simultaneous removal of bioaerosols, odors and volatile organic compounds from a wastewater treatment plant by a full-scale integrated reactor. *Process Saf. Environ. Prot.*, 144, 2–14.

Marquez, P., Benitez, A., Caballero, A., Siles, A., Martin, M.A. (2021). Integral evaluation of granular activated carbon at four stages of a full-scale WWTP deodorization system. *Sci. Total Environ.*, 754, 142237.

Martin, G. and Laffort, P. (1991). *Odeurs et désodorisation*. Lavoisier, Paris.

Martin, G. and Laffort, P. (1994). *Odors and Deodorization in the Environment*. VCH Publishing, New York.

Martin, G., Lemasle, M., Taha, S. (1992). Traitement des odeurs, bioépuration : innovations technologiques. *Techniques de l'Ingénieur*, 128–133.

Patria, L., Cathelain, M., Laurens, P., Barbère, J.P. (2001). Odour removal with a trickling filter at a small WWTP strongly influenced by tourism season. *Water Sci. Tech.*, 44(2–3), 243–249.

Piccardo, M.T., Geretto, M., Pulliero, A., Izzotti, A. (2022). Odor emissions: A public health concern for health risk perception. *Environ. Res.*, 204, 112121.

Rinker, E.B. and Sandall, O.C. (2000). Physical solubility of hydrogen sulfide in several aqueous solvents. *Can. J. Chem. Eng.*, 78, 232–236.

Ruiz-Munoz, A., Siles, J.A., Marquez, P., Toledo, M., Gutierrez, M.C., Martín, M.A. (2023). Odor emission assessment of different WWTPs with extended aeration activated sludge and rotating biological contactor technologies in the province of Cordoba (Spain). *J. Environ. Manag.*, 326(617), 116741–116752, Part A.

Talaiekhozani, M., Bagheri, A., Goli, M.R., Khoozani, T. (2016). An overview of principles of odor production, emission, and control methods in wastewater collection and treatment systems. *J. Environ. Manag.*, 170, 186–206.

Toledo, M. and Munoz, R. (2023). Odour prevention strategies in wastewater treatment plants: A pilot scale study of activated sludge recycling and oxidized nitrogen recycling. *J. Environ. Chem. Eng.*, 11(5), 110366–110376.

Zhang, L., De Schryver, P., De Gusseme, B., De Muynck, W., Boon, N., Verstraete, W. (2008). Chemical and biological technologies for hydrogen sulfide emission control in sewer systems: A review. *Water Res.*, 42, 1–12.

Index

E, F

I, L

M, N

O, P

R, S

T, W

Other titles from

ISTE

in

Ecological Science

2026

GAID Kader
Wastewater Engineering and Treatment 1: Composition, Pollutant Loads, Primary Settling, Activated Sludge
Wastewater Engineering and Treatment 2: Membranes, Hybrid Systems, Bacteria Beds and Biofilters, Biological Disks
Wastewater Engineering and Treatment 3: Elimination of Nitrogen and Phosphorus, Clarification

2025

COURANT Dimitri, REBER Bernard
Deliberative Democracy and Ecological Transition: The French Citizens' Convention for Climate
(Planetary Changes, Decision, Governance Set – Volume 1)

2024

AMIARD Jean-Claude, ZERBIB Jean-Claude
Radioactive Risk for Humans
(Radioactive Risk Set – Volume 7)

AKBAR Saima, KUMAR SINGH Suraj, KANGA Shruti
Disaster Management: Preparedness and Mitigation

FLEURENCE Joël
Aquatic Biotechnologies: From Genetic Engineering to Enzymatic or Fermentation Engineering

GAID Kader
Wastewater Reuse 1: Characteristics, Uses, Applications, Filtration and Disinfection of Water
Wastewater Reuse 2: Micropollutants, Membranes and Treatment Procedures

MATAGNE Patrick
Birth of Scientific Ecology: Eugenius Warming (1841–1924)

2023

GAID Kader
Drinking Water Treatment 1: Water Quality and Clarification
Drinking Water Treatment 2: Chemical and Physical Elimination of Organic Substances and Particles
Drinking Water Treatment 3: Organic and Mineral Micropolluants
Drinking Water Treatment 4: Membranes Applied to Drinking Water and Desalination
Drinking Water Treatment 5: Calco-carbonic Equilibrium and Disinfection

2022

AMIARD Jean-Claude
Marine Radioecology
(Radioactive Risk Set – Volume 6)

GODET Laurent, DUFOUR Simon, ROLLET Anne-Julia
The Baseline Concept in Biodiversity Conservation: Being Nostalgic or Not in the Anthropocene Era

GRISON Claude, CASES Lucie, LE MOIGNE Mailys, HOSSAERT-MCKEY Martine
Photovoltaism, Agriculture and Ecology: From Agrivoltaism to Ecovoltaism

LÉVÊQUE Christian
Biodiversity Erosion: Issues and Questions

2021

AMIARD Jean-Claude
Disarmament and Decommissioning in the Nuclear Domain (Radioactive Risk Set – Volume 4)
Management of Radioactive Waste (Radioactive Risk Set – Volume 5)

FLEURENCE Joël
Microalgae: From Future Food to Cellular Factory

JOLY Fernand, BOURRIÉ Guilhem
Mankind and Deserts 1: Deserts, Aridity, Exploration and Conquests
Mankind and Deserts 2: Water and Salts
Mankind and Deserts 3: Wind in Deserts and Civilizations

PIOCH Sylvain, SOUCHE Jean-Claude
Eco-design of Marine Infrastructures: Towards Ecologically-informed Coastal and Ocean Development

2020

BRUSLÉ Jacques, QUIGNARD Jean-Pierre
Fish Behavior 1: Eco-ethology
Fish Behavior 2: Ethophysiology

ROSSIGNOL Jean-Yves
Climatic Impact of Activities: Methodological Guide for Analysis and Action

2019

AMIARD Jean-Claude
Industrial and Medical Nuclear Accidents: Environmental, Ecological, Health and Socio-economic Consequences
(Radioactive Risk Set – Volume 2)
Nuclear Accidents: Prevention and Management of an Accidental Crisis
(Radioactive Risk Set – Volume 3)

BOULEAU Gabrielle
Politicization of Ecological Issues: From Environmental Forms to Environmental Motives

GIRAULT Yves
UNESCO Global Geoparks: Tension Between Territorial Development and Heritage Enhancement

KARA Mohamed Hichem, QUIGNARD Jean-Pierre
Fishes in Lagoons and Estuaries in the Mediterranean 2: Sedentary Fish
Fishes in Lagoons and Estuaries in the Mediterranean 3A: Migratory Fish
Fishes in Lagoons and Estuaries in the Mediterranean 3B: Migratory Fish

OUVRARD Benjamin, STENGER Anne
Incentives and Environmental Policies: From Theory to Empirical Novelties

2018

AMIARD Jean-Claude
Military Nuclear Accidents: Environmental, Ecological, Health and Socio-economic Consequences
(Radioactive Risk Set – Volume 1)

FLIPO Fabrice
The Coming Authoritarian Ecology

GUILLOUX Bleuenn
Marine Genetic Resources, R&D and the Law 1: Complex Objects of Use

KARA Mohamed Hichem, QUIGNARD Jean-Pierre
Fishes in Lagoons and Estuaries in the Mediterranean 1: Diversity, Bioecology and Exploitation

2016

BAGNÈRES Anne-Geneviève, HOSSAERT-MCKEY Martine
Chemical Ecology

2014

DE LARMINAT Philippe
Climate Change: Identification and Projections

Printed and bound by CPI Group (UK) Ltd, Croydon, CR0 4YY
11/08/2026
14939851-0001